# Ada and Beyond

## Software Policies
## for the Department
## of Defense

Committee on the Past and Present Contexts for the Use of Ada in the
Department of Defense

Computer Science and Telecommunications Board

Commission on Physical Sciences, Mathematics, and Applications

National Research Council

NATIONAL ACADEMY PRESS
Washington, D.C. 1997

NOTICE: The project that is the subject of this report was approved by the Governing Board of the National Research Council, whose members are drawn from the councils of the National Academy of Sciences, the National Academy of Engineering, and the Institute of Medicine. The members of the committee responsible for the report were chosen for their special competences and with regard for appropriate balance.

This report has been reviewed by a group other than the authors according to procedures approved by a Report Review Committee consisting of members of the National Academy of Sciences, the National Academy of Engineering, and the Institute of Medicine.

The National Academy of Sciences is a private, nonprofit, self-perpetuating society of distinguished scholars engaged in scientific and engineering research, dedicated to the furtherance of science and technology and to their use for the general welfare. Upon the authority of the charter granted to it by the Congress in 1863, the Academy has a mandate that requires it to advise the federal government on scientific and technical matters. Dr. Bruce Alberts is president of the National Academy of Sciences.

The National Academy of Engineering was established in 1964, under the charter of the National Academy of Sciences, as a parallel organization of outstanding engineers. It is autonomous in its administration and in the selection of its members, sharing with the National Academy of Sciences the responsibility for advising the federal government. The National Academy of Engineering also sponsors engineering programs aimed at meeting national needs, encourages education and research, and recognizes the superior achievements of engineers. Dr. William A. Wulf is interim president of the National Academy of Engineering.

The Institute of Medicine was established in 1970 by the National Academy of Sciences to secure the services of eminent members of appropriate professions in the examination of policy matters pertaining to the health of the public. The Institute acts under the responsibility given to the National Academy of Sciences by its congressional charter to be an adviser to the federal government and, upon its own initiative, to identify issues of medical care, research, and education. Dr. Kenneth I. Shine is president of the Institute of Medicine.

The National Research Council was organized by the National Academy of Sciences in 1916 to associate the broad community of science and technology with the Academy's purposes of furthering knowledge and advising the federal government. Functioning in accordance with general policies determined by the Academy, the Council has become the principal operating agency of both the National Academy of Sciences and the National Academy of Engineering in providing services to the government, the public, and the scientific and engineering communities. The Council is administered jointly by both Academies and the Institute of Medicine. Dr. Bruce Alberts and Dr. William A. Wulf are chairman and interim vice chairman, respectively, of the National Research Council.

Support for this project was provided by the Department of Defense (under contract number DASW01-96-C0028). The views, options, and findings contained in this report are those of the authors and should not be construed as an official Department of Defense position, policy, or decision, unless so designated by other official documentation.

Library of Congress Catalog Card Number 96-71960
International Standard Book Number 0-309-05597-0

Additional copies of this report are available from:

National Academy Press
2101 Constitution Avenue, NW
Box 285
Washington, DC 20055
800/624-6242
202/334-3313 (in the Washington Metropolitan Area)
http://www.nap.edu

# Preface

It is increasingly important for the Department of Defense (DOD) to implement effective information systems policies and strategies, as future battles will be decided as much in "cyberspace" as in physical space. The use of effective computer programming languages—and more broadly, of software engineering technology and policy designed for optimal support of DOD requirements—is key to DOD's strategy of achieving information dominance for warfighting. For the past two decades, DOD has used programming language policy as a vehicle for obtaining cost-effective, high-performance information systems. However, the process of software development has changed considerably during this period, as has the computer industry itself. These changes have altered the environment in which DOD develops and produces information systems.

It is in this context that Assistant Secretary of Defense (Command, Control, Communications, and Intelligence) Emmett Paige, Jr., requested that the National Research Council's Computer Science and Telecommunications Board (CSTB) review DOD's current programming language policy. Convened by CSTB, the Committee on the Past and Present Contexts for the Use of Ada in the Department of Defense was asked to:

1. Review DOD's original (mid-1970s) goals and strategy for the Ada program;
2. Compare and contrast the past and present environments for DOD software development; and
3. Consider alternatives and propose a refined set of goals, objectives, and approaches better suited to meeting DOD's software needs in the face of ongoing technological change.

Although the committee focused on programming language issues, it also considered them in the context of software architectures, components, and life-cycle processes, consistent with the realization that successful software engineering strategy involves several elements that are at least as important as the programming language component.

Throughout its deliberations, the committee was sensitive to the fact that the issues surrounding Ada and DOD programming language policy have been the source of vigorous debate among DOD policymakers, program managers, government contractors, and the software community at large.  Thus the committee made a concerted effort to collect a variety of views, and it received numerous briefings, position papers, and analyses from representatives of government agencies, as well as the defense, aerospace, and commercial industries.   The committee membership included many different organizational viewpoints and personal experiences; as it reflected the larger community, so also did it engage in vigorous debate during its own deliberations.   In the process of reaching conclusions and formulating recommendations, however, the committee agreed on the importance of DOD adopting software policies that better reflect ongoing significant changes in the discipline of software engineering, while retaining the benefits of prior investment and policy decisions.

The committee also understood the desire on all sides to bring closure to a policy debate that has continued for many years.  Several briefings to the committee included requests that the committee not suggest further studies on the topic.  The committee found these requests compelling and has attempted to frame its recommendations so that they can be acted on directly by DOD policymakers.  Thus, for example, in addition to making recommendations in the main text of the report concerning the appropriate scope of and criteria for DOD software policy, the committee found it useful to propose a revised statement of the current policy as embodied in DOD Directive 3405.1.  The committee-modified form of the DOD-revised draft (May 15, 1996) of the directive is offered in Appendix A for consideration as a template for further revision. The committee was aware that DOD has been conducting an effort to revise this policy; indeed, the committee was provided copies of two different draft revisions.

In addition to the individuals and organizations who participated in committee meetings and wrote position papers for the study (listed in Appendix E), the committee would like to acknowledge the numerous anonymous reviewers for their constructive comments on a draft version of this report.  The committee would also like to acknowledge the efforts of Assistant Secretary Paige and his staff, including Cynthia Rand and Connie Leonard, for assisting the committee in locating individuals and materials to consult.

Finally, the committee would like to acknowledge the support provided by the Computer Science and Telecommunications Board and staff.  Several Board members took an active interest in the project and offered numerous suggestions that helped to strengthen the report.   The CSTB staff were instrumental in organizing the committee meetings and coordinating briefings, reviews, and interactions with Board members.  In particular, CSTB's administrative assistant, Gloria Bemah, provided excellent administrative support, and its director, Marjory Blumenthal, played a key role in overseeing the study on behalf of the CSTB.  Susan Maurizi edited the report under a compressed schedule, and Gail Pritchard and Jean Smith of CSTB assisted in production of the final draft.  Finally, Paul Semenza, the study director, worked closely with the committee in every phase of the study.

# Contents

# Executive Summary

## WHAT SHOULD THE DEPARTMENT OF DEFENSE DO ABOUT THE ADA PROGRAMMING LANGUAGE?

Ada was developed in the 1970s by the Department of Defense (DOD) and adopted in some DOD programs until its use was required for new software development in 1987. It has been employed as a tool for developing quality software and as a DOD policy lever to encourage DOD organizations and programs to adopt modern software engineering principles. Changes within DOD, the software engineering community, and the commercial software industry have led DOD to reassess its programming language policy.

The Committee on the Past and Present Contexts for the Use of Ada in the Department of Defense was created by the National Research Council's Computer Science and Telecommunications Board to review DOD's programming language policy and the question of Ada's role in it. This study presents findings and recommendations developed by the committee for DOD's consideration in efforts to revise its current policy.

The committee concluded that a vigorous Ada program would enhance the reliability and performance of DOD warfighting systems, and it recommends that DOD continue the use and promotion of Ada in such systems. However, the committee found significant problems with the two primary components of DOD's current strategy for Ada. First, there are problems in the scope, design, and implementation of the current programming language policy, which requires the use of Ada for all software to be maintained by DOD; the committee recommends several modifications. Second, DOD's plan to discontinue investments in Ada technology and user-community support by the end of 1997 will weaken the Ada infrastructure and work against any requirement for DOD systems to use Ada in the future; given the large installed base of Ada code in warfighting systems, targeted investments in Ada are justified.

In the course of this study, the committee also concluded that the currently available data on effects of programming language on project outcomes are insufficient, on their own, to serve as a basis for strong determinations of the impact of programming language choice on the outcome of DOD

1

programs. Briefings received by the committee also highlighted the difficulty that DOD managers have in gaining access to data that can support informed decisions. The committee did find that trends in the data, anecdotal evidence, and expert judgment provided a basis for its finding that Ada provides benefits in warfighting systems. However, based on its experience with the limitations of currently available data, the committee makes an additional recommendation that DOD institute a corporate data collection effort and develop metrics as a sound basis for evaluating software so as to guide future policy and management decisions.

## CONTEXT AND TRENDS

DOD's policy preference for Ada had some merit a decade ago. Most software at the time was entirely custom, and Ada had a good track record in delivering custom software with higher quality and lower life-cycle costs. However, a custom Ada solution is no longer the best approach in many application areas, due to the following major trends:

- *COTS as a source of information infrastructure for applications.* Software solutions increasingly depend on commercial off-the-shelf (COTS) software, which provides much of an application's information infrastructure: operating system, database management, networking, user interface, and distributed processing functions. Much of this software is written in programming languages other than Ada that often do not have readily available interfaces to programs written in Ada. Developing these interfaces is not a major technical problem, but, particularly in the area of commercial Internet applications, COTS software is evolving rapidly, making it hard for Ada solutions to keep current.
- *Product-line solutions and production factors.* Software for many application areas is achieving economies of scale through the development of product-line architectures, enabling software assets to be reused across families of applications. These product-line solutions are driven by strongly coupled "production factors," including software components, processes and methods, human resources, and expertise in particular domains. In warfighting application areas such as weapon control and electronic warfare, there is little commercial development, and DOD has established a strong community of warfighting software developers whose production factors are oriented to Ada. However, for the numerous DOD applications in which the market is dominated by commercial solutions, such as finance and logistics, production factors have been built around programming languages other than Ada, putting Ada solutions at a disadvantage.

Additional conditions that strongly influenced the committee's findings and recommendations include the following:

- *DOD emphasis on achieving information dominance.* According to Secretary of Defense William Perry, ". . . our warfighting strategy sustains and builds on . . . the application of information technology to gain great military leverage to continue to give us [an] unfair competitive advantage" (Perry, 1996a). This assertion highlights the importance of a capability for enhancing military competitive advantage as a criterion for the choice of programming language.
- *Large and increasing inventory of DOD Ada software.* DOD now has over 50 million lines of operational Ada weapon systems software, with a great deal more under development. Most of this software is in critical warfighting application areas, and there are no quick and cheap ways to translate it into other languages. DOD policies and investment strategies that weaken Ada support for this software are very risky because of the role warfighting software plays in maintaining national security.

- *Proliferation of programming languages has decreased, but polylingualism is here to stay.* One goal of developing Ada was to reduce the proliferation of programming languages used in DOD systems, estimated to be approximately 450 in the 1970s. The number of languages used throughout DOD has indeed decreased: the use of machine and assembly languages has diminished, and the number of third-generation languages in use has been reduced. However, there has been a rapid increase in development of fourth-generation languages by the commercial sector (there are now more than 100 different such languages), and use of these languages by DOD is increasing. Thus, DOD cannot expect to avoid polylingual software solutions. However, support for multilanguage applications has improved significantly.

- *Programming language choice is one of several key software engineering decisions.* The requirement to use Ada and the process for obtaining a waiver isolate programming language decisions from other key software engineering decisions (e.g., choices of computer and software architectures, decisions about use of COTS components, and milestone schedules). These decisions are also currently made at the system level, rather than at the component or subsystem level. This arrangement creates an incentive for DOD programs to make decisions that are not optimal for DOD as an organization. Future programming language decisions need to be made as part of an integrated software engineering process.

## FINDINGS AND RECOMMENDATIONS

The committee developed the following set of findings and recommendations for future DOD software policy and strategy. The recommendations address the use of Ada in warfighting software, the application in which the committee finds Ada to have demonstrated benefit; the proper scope and implementation of software policy; investment in Ada; and collection of data as a basis for assessing the effectiveness of software and software policy.

### Ada Competitive Advantage

**Finding**. Ada gives DOD a competitive advantage in warfighting software applications, including weapon control, electronic warfare, performance-critical surveillance, and battle management.

**Recommendation**. Continue vigorous promotion of Ada in warfighting application areas.

**Rationale**. Available project data and analyses of programming language features indicate that, compared with other programming languages, Ada provides DOD with higher-quality warfighting software at a lower life-cycle cost. DOD can increase its advantage by strengthening its Ada-based production factors (involving software tools, technology, and personnel) for warfighting software (see Chapters 2 and 3).

### Applicability of Policy to DOD Domains

**Finding**. DOD's current requirement for use of Ada is overly broad in its application to all DOD-maintained software.

**Recommendation**. Focus the Ada requirement on warfighting applications, particularly critical, real-time applications, in which Ada has demonstrated success. For commercially dominated applications, such as office and management support, routine operations support, asset monitoring, logistics, and medicine, the option of using Ada should be analyzed but should not be assumed to be preferable.

**Rationale**.  For warfighting software, supporting Ada-based production factors (involving software tools, technology, and personnel) gives DOD a competitive advantage.  In this domain, eliminating the use of Ada would both compromise this advantage and diminish the capabilities for maintaining DOD's existing 50 million lines of Ada.  In commercially dominated areas, pushing applications toward Ada would create a disadvantage for DOD (see Chapters 2 and 3).

## Scope of Policy

**Finding**.  DOD's current requirement for use of Ada overemphasizes programming language considerations.

**Recommendation**.  Broaden the current policy to integrate the choice of programming language with other key software engineering concerns, such as software requirements, architecture, process, and quality factors.

**Rationale**.  The current policy isolates the Ada requirement and the waiver process from other software engineering decisions, causing programs to make premature or non-optimal decisions (see Chapter 1).  DOD has already taken steps to broaden the policy focus in its draft revision of its programming language policy, DOD Directive 3405.1; this report recommends modifications to that draft policy (Appendix A).

## Policy Implementation

**Finding**.  DOD's current Ada requirement and the related waiver process have been weakly implemented.  Many programs have simply ignored the waiver process.  Other programs make programming language decisions at the system level, but often a mix of Ada and non-Ada subsystems is more appropriate (see Chapter 1).

**Recommendation**.  Integrate the Ada decision process with an overall Software Engineering Plan Review (SEPR) process.  Passing such a review should be a requirement for entering the system acquisition Milestone I and II reviews covered by DOD Instruction 5000.2.  It should also be required for systems not covered in 5000.2, and recommended by DOD for DOD-directed software development and maintenance of all kinds.

**Rationale**.  The SEPR concept is based on the highly successful commercial architecture review board practice.  The SEPR process involves peer reviewing not only the software and system development plans, but also the software and system architecture (building plan) and its ability to satisfy mission requirements, operational concepts, conformance with architectural frameworks, and budget and schedule constraints; the process also involves reviewing other key decisions such as choice of programming language (see Chapter 4).

## Investment in Ada

**Finding**.  For Ada to remain the strongest programming language for warfighting software, DOD must provide technology and infrastructure support.

**Recommendation**.  Invest in a significant level of support for Ada, or drop the Ada requirement. The strategy developed by the committee recommends an investment level of approximately $15 million per year.

**Rationale**.  With investment, DOD can create a significant Ada-based complex of production factors (involving software tools, technology, and personnel) for warfighting application domains.

Without such support, Ada will become a second-tier, niche language such as Jovial or CMS-2 (see Chapter 5).

## Software Metrics Data

**Finding**. DOD's incomplete and incommensurable base of software metrics data weakens its ability to make effective software policy, management, and technical decisions.

**Recommendation**. Establish a sustained commitment to collect and analyze consistent software metrics data.

**Rationale**. The five sets of findings and recommendations above are based on a mix of incomplete and incommensurable data, anecdotal evidence, and expert judgment. For this study, the patterns of consistency in these sources of evidence provide reasonable support for the results—but not as much as could be provided by quantitative analysis based on solid data. A few organizations within DOD have benefited significantly from efforts to provide a sound basis for software metrics; a DOD-wide data collection effort would magnify the net benefits (see Chapter 2).

## WHAT THE DEPARTMENT OF DEFENSE SHOULD DO ABOUT ADA

In summary, the committee concluded the following regarding DOD and Ada:

1.   DOD should continue to require Ada for its warfighting software and should drop the Ada requirement for its other software.

2.   DOD should provide roughly $15 million per year for Ada infrastructure support, or drop the requirement to use Ada entirely.

3.   DOD should make programming language decisions in the context of a Software Engineering Plan Review process.

The rationale for the above statements is as follows:

1.   In commercially dominated areas, although Ada may offer some advantages for custom software development, the preponderance of existing commercial activity and solutions in other languages counters these advantages, thereby shifting the business case away from mandating Ada in these areas.

2.   In warfighting applications, Ada's technical capabilities for building real-time, high-assurance custom software are generally superior to those of other programming languages. DOD's investments in Ada to date have provided DOD systems with a competitive advantage in these areas.

3.   The commercial marketplace alone will not sustain a robust Ada infrastructure.

4.   A relatively modest ($15 million per year) DOD investment at the margin would be sufficient to sustain a robust Ada infrastructure for warfighting applications.

5.   DOD's inventory of more than 50 million lines of Ada warfighting software will become a liability without a robust Ada infrastructure.

6.   DOD's current Ada waiver procedure can be effectively replaced by adoption of the commercially established practice of using architecture review boards, a process that can also strengthen DOD's overall software engineering capability.

**ORGANIZATION OF THIS REPORT**

Chapter 1 describes the past and present contexts for Ada and DOD software development and programming language policy and points out problems with the current approach. Chapter 2 relates issues in decisions about programming language to other software engineering decision issues and evaluates Ada's support for software engineering processes. It also summarizes the results of comparing the relative cost-effectiveness of Ada and other programming languages, based on analyses of language features and empirical data from software projects.

Chapter 3 presents a business-case analysis for the use of Ada within DOD, gives the committee's findings and recommendations, and analyzes them along with alternatives.

The software policy recommended by the committee for DOD is elaborated in Chapter 4, which also provides additional details on programming language selection criteria and explains the basic elements of the Software Engineering Plan Review process recommended by the committee. Chapter 5 discusses the committee's recommended strategy for investment in infrastructure for Ada.

Appendix A reproduces a draft revision of DOD Directive 3405.1 on programming language policy (DOD, 1987a), a document to which the committee has added further suggested modifications that reflect its recommendations in this report. Appendix B presents detailed descriptions of Ada and other programming languages; a glossary of terms used in the report is contained in Appendix C; Appendix D compares Ada's features with those of other third-generation programming languages; and Appendix E lists briefings and position papers received by the committee during the course of its study.

# 1

# The Changing Context for DOD Software Development

For nearly two decades, the Ada programming language has been a cornerstone of efforts by the Department of Defense (DOD) to improve its software engineering practices. DOD created Ada in the 1970s to serve as a department-wide standard that would satisfy its special requirements for embedded and mission-critical software, and would also encourage good software engineering. Both the new language and the new software engineering ideas associated with it met with some criticism, and both have evolved as a result. Today, Ada is the most commonly used language for mission-critical defense software, which includes weapon systems and performance-critical command, control, communications, and intelligence ($C^3I$) systems. DOD's inventory contains nearly 50 million lines of Ada code in these applications (Hook et al., 1995). Given the long operational life of such systems, DOD has made a significant investment in Ada technology. Ada is the second most commonly used language (after Cobol) for DOD automated information systems, which include payroll and logistics programs. The DOD inventory contains more than 8 million lines of Ada code in these applications (Hook et al., 1995).

Hopes for broad commercial adoption of Ada have not been realized, however. Its commercial use has been eclipsed by other languages, such as C, then C++, and, most recently, Java. DOD's inclusive approach in the development of the language, as well as its promotional campaigns in support of Ada, do not appear to have been successful in fostering adoption of the language beyond defense and other mission-critical applications.

During Ada's lifetime, DOD's position in the software market has shifted. Although DOD still has an influence, its share of the market has diminished—not because DOD's need for software has decreased, but rather because the size of the commercial software market has exploded, generating a corresponding increase in investments in commercial software technology. DOD made significant investments to develop Ada (both Ada 83 and Ada 95)[1] and mandated its use on certain DOD projects. The DOD requirement to use Ada appears to have been beneficial for custom software that has no commercial counterparts (e.g., weapon systems and performance-critical $C^3I$ software). On the other hand, this policy has frequently been counterproductive in application areas that have strong commercial support. In these areas, DOD's policy has inhibited DOD from taking advantage of existing commercial infrastructure written in or for other languages.

## GROWTH IN THE COMMERCIAL SOFTWARE INDUSTRY

Commercial software includes a great deal of powerful infrastructure software such as development tools, operating systems, database management systems, networking software, user interfaces, and transaction processsing programs.  It also includes rich and growing sources of software for applications that are similar to some DOD applications, such as management information systems; geographic information systems; and logistics, medical, engineering and scientific, and office-support systems.  With the exception of some aerospace, transportation, and safety-critical applications software, little of this commercial software is written in Ada.

In the 1990s, the computing field has been transformed by technological advances, particularly in networking and in low-cost personal computing with associated tools.  While these advances have had relatively little impact on traditional real-time embedded systems, they have completely altered the character of commercial information systems and the processes used to develop them.  Information systems are now commonly built with a two- or three-level client-server architecture, and with a graphical user interface that is logically separated from computational steps and from a relational database.  Specialized tools and fourth-generation programming languages (4GLs; see glossary, Appendix C) have been developed for building this class of applications.  Such tools and languages, exemplified by Visual Basic and PowerBuilder, are extremely efficient for building small and medium-sized applications, particularly where the demands for reliability and availability are less stringent than those for real-time embedded systems.  Similar tools are now becoming available for the deployment of information systems across organizational intranets and the World Wide Web.  Because in certain domains these tools and languages operate at a higher level than does any traditional programming language (including Ada), they are often the most appropriate way to prototype and develop information systems.  Finally, growth in Internet-based software has increased the already rapid pace of product development in the commercial software industry.

## OBSTACLES TO BROAD ADOPTION OF ADA

One goal of the inclusive and extensive process undertaken to develop Ada was to create a language that would be widely adopted by the software community, beyond DOD.  Utilizing commercial technology has become more important to DOD in recent years, as a combination of declining financial resources for DOD and great strides in commercial developments across all areas of advanced technology has led to an increasing emphasis on leveraging commercial technology in developing defense systems (DOD, 1994b, 1995b).

Ada has taken its place among the better known and widely used third-generation programming languages (3GLs; see glossary, Appendix C); however, it has not become as popular as its proponents had hoped.  One study of programming language use estimated that Ada 83 applications constitute only 2 percent of all computer applications in U.S. inventories, and slightly more than 3 percent of all function points (Jones, 1996b).  Ada is used primarily within the DOD community.  Beyond that community, it has been adopted by some software developers for the civilian market, especially where there is potential defense market cross-over or where there are similar requirements, such as in commercial aviation, process control, and medical instrumentation.[2]  However, this commercial use is a small fraction of the total commercial software market.

Another indicator of Ada's limited market penetration is the supply of and demand for Ada-trained programmers.  Jones (1996b) estimates that of the 1.92 million professional programmers in the United States, 90,000, or less than 5 percent, are Ada 83 programmers.[3]  In an informal review of software engineering employment opportunities advertised in two major newspapers, the committee noted that of more than 1,000 references made to individual programming languages and tools,[4] fewer than 3 percent of the citations referred to Ada; in comparison, C and C++ each accounted for more than 23 percent of the references.[5]   While there are many ways to look at Ada's current market share, employment opportunities and professional growth were recurring concerns expressed in many of the presentations made to the committee.

Given current conditions and probable trends, it is unlikely that the use of Ada, including the recent Ada 95 version, will grow significantly beyond the DOD-dominated and related commercial niche of high-assurance, real-time systems where it is already strong.  Some of the principal reasons for this conclusion are discussed in the following sections.

## Low Commercial Awareness and Limited Sponsorship

Ada has never attained the broad following associated with languages such as C++ and, most recently, Java.  Market research indicates that nearly all programming language decision makers in non-defense industries are aware of Cobol, C, C++, Fortran, and Pascal, but only two-thirds are aware of Ada; only one-fifth are familiar with Ada's characteristics (Telos, 1994).[6]

In decisions affecting adoption of programming languages, non-technical factors often dominate specific technical features.  These factors include the broad availability of inexpensive compilers and related tools for a wide variety of computing environments, as well as the availability of texts and related training materials.  In addition, grass-roots advocacy by an enthusiastic group of early users, especially in educational and research institutions, often has broad influence on adoption of programming languages.  These advantages were largely absent when Ada was introduced in 1980.  In contrast, C++ and Java both have achieved widespread acceptance and use.  The strong military orientation of the publicity generated for Ada also may have served to alienate significant portions of the academic and research communities.

Historically, DOD has been Ada's only sponsor, and Ada has been focused almost exclusively on the military niche occupied by DOD and its contractors.  In the past, Ada technology was subject to export control restrictions.  The development of tools and components funded by DOD was targeted to DOD organizations and defense contractors.  People and institutions outside DOD who were interested in Ada found it difficult to acquire compilers and training resources.

Many technical organizations evaluated Ada and its associated tools in the mid-1980s and decided not to use it.  Most of those organizations have committed significant resources to other languages and technologies; thus they are unwilling to reconsider Ada, even though Ada 95 is significantly different from Ada 83 and the tools are far more advanced.  Beyond the DOD-dominated niche, some organizations are unwilling to reconsider Ada because they continue to view Ada as a language that is suitable only for military applications.

Recently, DOD's Ada Joint Program Office has begun to promote the academic use of Ada by awarding educational grants and making lower-priced compilers available.  While these activities have had an impact (see the next section), by themselves they are unlikely to be enough to make Ada popular.  They have not been matched by development of infrastructure to make it attractive to the research community, where advanced software development is carried out and graduate students are trained.

## Limited Extent of Academic Instruction in Ada

The popularity of a programming language in the academic world and its use in industry are often linked.  Schools feel pressure to teach the languages that appear to be demanded most by the labor market.  At the same time, the adoption of computer languages in the classroom can lead to commercial use.[7]  A manager who has been exposed to a language in school is more likely to be confident about trying it out, and even more confident if it is one that many recent graduates appear to know.  University research and development groups are also influential, since the advanced software concepts they develop tend to influence the next generation of commercial products.  A language that is widely taught is more likely to be widely adopted.

Ada has not been widely taught in colleges and universities, particularly compared with Pascal, C, and C++; until recently, even the military academies taught programming in languages other than Ada.  A survey of more then 2,300 colleges and universities worldwide that have computer science curricula identified only 285 institutions that offer any courses in Ada; 237 of them are in the United States (IITRI, 1996), and they include many small institutions, among them community colleges and technical institutes.  Few of the leading computer science programs in the United States, as ranked by the National Research Council's assessment of graduate research programs (NRC, 1995), provide instruction in Ada:  only 6 of the top 53 programs (and 1 of the top 10) were identified by the 1996 survey as offering Ada courses.

However, the IIT Research Institute's 1996 survey of Ada instruction also found that in fewer than 3 years, there was a 47 percent increase in institutions offering courses in Ada and a 43 percent increase in the number of Ada courses offered (both measured worldwide).  A survey of institutions adopting Ada as a foundation language found that from 1993 to 1996 the number increased from 57 to 147 (Feldman, 1996).[8]  Yet the total is still comparatively small, and it is unclear how long this trend might continue without the strong sponsorship provided by DOD.

### Limited Availability of Ada Tools and Compilers

Historically, compilers and other language-specific tools for Ada have been significantly more costly and slower in coming to market than those for C and C++.  Initially, this was a matter of technology.  Since Ada embodied new technology that provided technical challenges for compiler writers, production-quality Ada compilers were not available for several years after Ada's official debut in 1980.  When they became available, they were large, slow, unreliable, and expensive.  This slowed DOD contractors' transition to Ada from other languages and soured some early users.  Because C compilers are much easier to build and have a higher expected sales volume, they have typically been the first available compilers for new microprocessors; C++ compilers have usually followed, and Ada compilers have often been the last to be available.  The problems with Ada compilers also impeded the use of Ada in education.  Thus, organizations seeking to adopt Ada faced near-term costs for new tools, especially the high-priced compilers, in addition to the cost of training people in a language that was not widely taught in academic institutions.

Currently, Ada compilers are comparable in quality to those of other 3GLs, and the increased hardware resources needed to run popular software, such as Windows 95, make the requirements of Ada compilers appear more modest.  In addition, the availability of the GNU NYU Ada 95 Translator (GNAT) has reduced the cost and improved the availability of Ada compilers.  GNAT is a component of the GNU compiler suite, sharing code-generation facilities with the GNU C and C++ compilers.  The GNU C compiler is generally recognized as a high-quality compiler.  The GNU technology makes it comparatively easy to support new processors; therefore, the GNU compiler is likely to be one of the first available when a new processor appears.  A non-technical but important side benefit is the association of GNAT with the Free Software Foundation, which should help GNAT to shed some of Ada's military-only stereotype.  The entire GNU compiler suite is distributed widely over the Internet, without charge, and is also distributed by some hardware vendors.  In addition to the freely available GNAT, the main compiler vendors (see below) also offer academic compilers to students at reduced prices.

The market for Ada compilers and tools has been estimated at $200 million annually.[9]  The modest size of this market has resulted in significant consolidation of Ada compiler and tool vendors over the past few years.[10]  There are now two dominant suppliers: Rational Software Corporation (which was formed by the merger of Rational with Verdix, which had previously acquired Meridian) and Aonix (previously Thomson Software Products, which acquired Alsys and Telesoft).  In addition, there are several other vendors that focus mainly on niche markets.  These suppliers include the Texas Instruments unit that was previously Tartan, TLD, OC Systems, DDC-I, RR Software, Intermetrics, Green Hills, ICC, and Ada Core Technologies (which was formed to support and commercialize GNAT).  Several of the vendors, including the two largest, have product lines other than Ada.  It is not clear how this consolidation will affect the availability and price of commercial Ada compilers in the long term.

## Assumption That Ada Has to Control Everything

Ada provides some services, such as input/output, multitasking, time keeping, and interrupt handling, that are traditionally in the domain of an operating system.  Early Ada implementations were designed with the assumption that Ada was in complete control of the hardware, with no operating system, and that all the software on a machine was written in Ada.  This approach is effective for the programming of embedded systems, in which the applications need to run without an operating system.  Since Ada hides differences between operating systems, these characteristics make applications more portable.

However, these features also mean that, when compared with a language in which services are obtained by operating system calls, it is harder to write an Ada run-time system.  It is also more difficult to port it to a new operating system, although this cost can be balanced by the benefit of reduced effort in writing and porting Ada application programs.  In early Ada applications, when compiler vendors lacked expertise in real-time kernel building, the run-time systems were a frequent cause of complaints related to the closed nature of the interface.  Developers of embedded systems applications wanted to have more direct control over the hardware resources.  Developers of conventional operating system applications were hampered by the lack of access to the full operating system functionality, and by the incompatibility of libraries written in other languages with the Ada run-time system.  The situation has improved in recent years; Ada vendors have adopted a more "open systems" approach in which the Ada run-time system is layered over a commercial real-time kernel or a traditional operating system, and Ada 95 supports interfaces to other languages.

## Need for Ada-compatible Application Programming Interfaces

An application programming interface (API) is a set of procedure and function specifications that provides access to the capabilities of a reusable software component, such as an operating subsystem for "windows" or network communications.  The vendors of most commercial off-the-shelf (COTS) software components typically provide a C-language API.  For the COTS component to be usable by an Ada application, an Ada-compatible API must be provided; this does not mean, however, that the COTS product itself needs to be written in Ada (a misconception that was evident in several presentations to the committee).  A vendor-provided Ada API often lags the C version by months or years (a very long time in the computer industry), and often costs more.  An Ada developer can create an interface to a C-language API without vendor support, but doing so can require intimate knowledge of the particular COTS product and/or the Ada language implementation.  The earlier implementation of non-Ada APIs, and greater vendor involvement, also have led to earlier standardization.  Thus, the time delay and extra cost or effort of obtaining an Ada API, and the delay in standardization, have become disincentives for the use of Ada.  An added disincentive is the challenge of keeping Ada APIs current with the frequent changes in COTS product features.

## Labor Market Forces

Software engineers are likely to be interested in enhancing skills that they expect to be most valuable in the software engineering marketplace, which is now dominated by commercial opportunities. Thus, programmers have moved quickly to learn Java and Hypertext Markup Language (HTML; used to develop pages for the World Wide Web) because they see these as the next wave, which can carry them to new career opportunities.  Similarly, software engineers might avoid using Ada if they see it as limiting their careers.  Given the cyclical nature of DOD spending, recent downsizing, and layoffs, defense software engineers have reasons to consider preparing for a move to the commercial sector. Even those who believe Ada is technically the best solution for a given program may face conflicting incentives in choosing a programming language.

On the other hand, the committee heard testimony that, for developing the specialized software that is most critical to the DOD mission, knowledge of the application domain is harder to obtain and more valuable than knowledge of a particular programming language, or even of software engineering itself.  Thus, software engineers who have expertise in defense-oriented applications are likely to be in greater demand in that sector than in the commercial marketplace, where their domain-specific skills would be less applicable.  Likewise, employers in the DOD sector are highly motivated to keep experienced engineers because of the expense of training new ones in the relevant applications, as well as the cost and delay of obtaining a security clearance for a person entering from the commercial market. These dynamics contribute to the separation of the military and commercial markets for software engineers, which is similar to the separation of those same markets for aerospace engineers.

## DOD PROGRAMMING LANGUAGE POLICY

### Policy History

DOD's decision to design Ada as a new programming language for embedded applications was a reaction to both the "software crisis" of the late 1960s and early 1970s and the advent of software engineering concepts.  It was also a response to the fact that each of the military services had developed separate programming languages that each was planning to independently standardize, upgrade, and improve.  Within DOD, software problems were contributing to project cost overruns and lengthy delays in system deployment.  From 1968 to 1973, the total cost of DOD's computer systems increased by 51 percent, even though hardware costs were decreasing dramatically (Fisher, 1976).  While some have argued that this increase could be interpreted as the legitimate cost of obtaining new functionality, at the time it was viewed as symptomatic of software development problems.

One visible aspect of DOD's software crisis was that systems were being developed in many different programming languages for many different computers, diluting resources and increasing the cost and complexity of maintenance.  Many of the systems were written in assembly language for specialized, proprietary processors, or written in programming languages unique to a particular project or contractor.  There was never a thorough count of the number of languages in use, but a widely cited estimate is "at least 450 general-purpose languages and dialects" (Hook et al., 1995).  The abundance of languages made it uneconomical to develop high-quality software tools and was an obstacle to using programmers and software across projects.  It was also a major source of post-deployment problems in areas such as interoperability, operations, and maintenance, and it hindered effective product-line management. Maintenance of compilers, assemblers, linkers, and other tools for all these languages was a significant burden, as were the hiring and training of programmers. In many cases, a system could be maintained only by its original developer; such vendor "lock-in" added to maintenance expenses.

For these reasons, it appeared desirable for DOD to converge on a minimal set of programming languages. Representatives of various DOD and allied defense organizations worked together to define the DOD requirements for high-order languages. An early outcome was the release of DOD Directive 5000.29 (DOD, 1976). The directive required the use of an approved high-order language, unless another choice could be shown to be more cost-effective, and it established a single point of control for each language. However, it was determined that none of the existing languages met the requirements for embedded systems, which were estimated to represent more than half of all DOD software costs in 1973 (Fisher, 1974).

DOD decided to design a new language that would serve as the single, common, high-order language. While DOD had other software problems that went deeper than those associated with programming languages, it appeared that programming language problems were amenable to a technical solution. Moreover, the conversion to a new, common, high-order programming language was viewed by some as a vehicle for DOD-wide efforts to improve software engineering. The new DOD language, which eventually became Ada, was intended to be a modern programming language that would reflect the accumulated knowledge of programming language design and provide the appropriate set of concepts and features for implementing embedded systems.

Representatives of key stakeholders, including organizations within DOD, its contractors, and many of the world's software engineering and programming language experts, were involved in the identification of requirements, the design and evaluation of the early prototype languages, and the refinement of the preliminary Ada design. The end product, eventually named Ada 83, was officially released in 1980 and became a standard in 1983. It was recognized as a powerful, modern programming language that addressed DOD's stated requirements for embedded systems. However, Ada's adoption within DOD and by its contractors did not proceed as quickly as anticipated.

In 1983, Undersecretary of Defense Richard DeLauer established a policy mandating the use of Ada for all new mission-critical applications. The policy specified a waiver process, whereby other DOD authorized languages (CMS-2M/Y, Jovial J3/J73, Fortran, Cobol, TACPOL, SPL/1, and C/ATLAS) could also be used. While the preferential treatment for Ada helped, Ada support software was not yet mature, and there were still many contract awards across the services where Ada was not selected. As a result, systems developed in Ada continued to be in the minority. That situation led to the establishment, in 1987, of the current policy, specified by DOD Directive 3405.1 (DOD, 1987a), which prescribes the use of Ada for most DOD procurements and internal developments. The original list of approved high-order languages was expanded to include Minimal BASIC and Pascal.

The proliferation of programming languages within DOD systems, one of the original reasons for mandating the use of Ada, appears to have diminished over time. Most new DOD software is written in Ada or one of a few other languages. As of 1994, 79 percent of all DOD mission-critical software was written in 3GLs; of this code, 33 percent was written in Ada, 37 percent in the other approved high-order languages, 22 percent in C, and 3 percent in C++ (Hook et al., 1995). It is difficult to tell how much of this consolidation of language use is a result of the requirement to use DOD-approved languages, given that C—the second most widely used language in DOD—has never been on DOD's list of approved languages. Certainly larger market and industry forces have also been at work. For example, growing standardization in the computer industry has resulted in fewer new computer architectures being introduced, particularly when compared with 20 years ago, resulting in fewer assembly languages in use. Similarly, the rate of growth in new 3GLs has diminished since the 1970s and has been overtaken by development of the infrastructure and culture needed to build software involving components in different programming languages.

However, the growth of 4GLs indicates a potential for a new proliferation of programming languages. For example, one study lists 115 languages (out of a total of 455 currently active languages) that meet one definition of a 4GL, i.e., that the language requires 30 or fewer statements to encode one function point (Jones, 1996c). 4GLs typically are not intended to serve as general-purpose programming languages, but they may include a lower-level sub-language that is more general and allows users to program functionality that is not built into the 4GL. Hook et al. (1995) found that 4GLs are used increasingly for DOD automated information systems applications; the committee also heard from DOD representatives that 4GLs increasingly are being used for rapid-development special applications. Thus, it appears that DOD will continue to operate in a "polylingual" world.[11]

## Ada's Place in Current DOD Programming Language Policy

DOD policy states that Ada is to be the "single, common, computer programming language for Defense computer resources used in intelligence systems, for the command and control of military forces, or as an integral part of a weapon system" (DOD, 1987a). The policy allows for the use of other, previously authorized languages in deployment and maintenance, but not for redesign or addition of more than one-third of the software. Ada is to be used "for all other applications, except when the use of another approved higher order language [HOL] is more cost-effective over the application's life-cycle." It ranks as preferences (1) off-the-shelf applications packages and advanced software technology, (2) Ada-based software and tools, and (3) approved standard HOLs. Exceptions are allowed only by the granting of a waiver, which requires that the alternative language be more cost-effective and that it be chosen from the list of DOD-approved languages. Neither Ada use nor a waiver is required for COTS or contractor-maintained software developments or for vendor-provided updates.

## Implementation of Policy on Waivers

Requests for waivers to develop systems in different languages are handled at a very high management level—the offices of the Assistant Secretary ($C^3I$) or the Service Acquisition Executives— and are reviewed independently from the requesting program's other key decision milestones (such as the Defense Acquisition Board and Major Automated Information Systems Review Council). In practice, such waivers have rarely been requested. The committee was informed that 31 waivers had been granted since 1992 across the services (3 by the Army, and 14 each by the Navy and the Air Force). Because most requests for a waiver have been granted, this relatively small number of approved waivers suggests that only a very small percentage of the many projects that did not use Ada actually submitted a waiver request.[12]

Based on briefings and testimony to the committee and the information discussed above, the committee concluded the following about the implementation and some of the effects of the Ada waiver policy:

1. Many projects have ignored or manipulated the policy on waivers, employing languages other than Ada without the required waiver.

2. Many project managers fear that requesting a waiver will reflect badly on them; this has caused some to employ Ada where it is not cost-effective.

3. The DOD and services' authorities generally have the capability to grant waivers that are justified, given the small number of waiver requests.

4.  The granting authorities do not have the capability and expertise to evaluate the technical merits and long-term consequences of the full number of Ada waiver requests that would be made if there were full compliance with the policy.

In organizations with a high level of understanding of software, similar waiver processes can work reasonably well. Waivers are requested by developers where they make sense; waivers are granted by managers where they make sense; and the developers and managers know enough about software to reconcile differences of opinion on what makes sense. One of the recurring themes in successful DOD projects using Ada was that Ada was selected for appropriate technical and economic reasons. However, selection of Ada solely to satisfy DOD's overall policy on programming languages has not been a guarantee of project success.

## Importance of Appropriate Expertise

Custodians of mandates on software use who do not possess sufficient knowledge of software tend to rely too much on narrow interpretation of the mandates, and DOD historically has not had a high level of software expertise. The Defense Science Board found in 1987 that DOD lacked adequate career paths for software professionals and had long ignored its software personnel problems (DOD, 1987b). Testimony heard by this committee indicated that although the level of software training has increased, such problems persist. For example, cutbacks in several DOD organizations in the early 1990s appear to have caused numerous software experts to leave DOD for industry. The approach that the Defense Science Board recommended was, "Do not believe DOD can solve its skilled personnel shortage; plan how best to live with it, and how to ameliorate it."

## Level of Applicability

Another recurring issue and ambiguity in current DOD policy on programming languages is that it reflects a system-level view of software that does not consider subsystems independently. For example, for a project with three subsystems—(1) an operational flight program (all new software), (2) a simulator (based on an existing simulator written in C), and (3) a ground-based test capability (combining new and legacy components in multiple languages)—current DOD policy encourages project managers to either write all subsystems in Ada or apply for a waiver for all subsystems. This approach leads to a simplistic choice between two options, neither of which is optimal. A clear, more flexible policy is needed that allows project managers to optimize programming language use at the subsystem and component level, without incurring a penalty of additional administrative overhead for the division into components.

## Implications

To be effective, DOD's policy requiring the use of Ada must include positive incentives for doing so, and it must be implemented closer to the project level within DOD. The current policy fails on both accounts. It has often had negative effects on DOD software engineering processes, in particular because DOD's policy on waivers for use of alternative programming languages has been implemented unevenly by DOD staff who lack the necessary technical knowledge, understanding of the relevant details of system design, or the motivation to consider long-term and service-wide objectives. Many DOD personnel testified to the committee that waivers are perceived as difficult to defend (even though it appears that most requests are actually granted).  This perception frequently has led to manipulation, bypassing, or simply ignoring of the waiver process.  Narrow interpretation of the policy has led to a number of poor decisions to use Ada, even when other solutions offered significant improvements in capability. For example, certain graphical user interface development tools have frequently not been used simply because they did not generate Ada or were not written in Ada.

## DOD INVESTMENT STRATEGY

DOD weapon systems programs and commercial organizations both understand that significant post-development investments are needed to keep software systems functioning effectively.  For example, Citibank spends 80 percent of its software budget on sustaining and enhancing existing code.[13] In the programming language area, Eric Schmidt, chief technology officer at Sun Microsystems, has stated that "several years" lie between Java and black ink (Aley, 1996).  It is reasonable to assume that Ada 95 will also require ongoing investment.

DOD has assigned the responsibility for sustaining Ada to the Defense Information Systems Agency (DISA), under the Assistant Secretary of Defense ($C^3I$).  It is the committee's understanding that DISA's current plan is to shrink the originally planned budget of $10 million in annual support for Ada 95 to nearly zero by the end of 1997.  The committee does not believe that Ada 95 is an exception to the general rule that software requires continuing investment to remain effective; briefings from DISA indicated that it has made an assumption to the effect that "the language exists and is mature," meaning that the commercial sector will provide support.  The committee disagrees with this assessment.

The barriers to commercial adoption of Ada discussed above in this chapter are a significant concern, because without support and promotion by critical customers such as DOD and the commercial safety-critical community, there is a serious danger that the Ada tool and compiler industry will shrink to the point that it can no longer provide widespread support to warfighting systems.  This will ultimately increase DOD's costs, because it will have to take over full maintenance and development  for Ada tools.  DOD may also have to use programming languages that will result in more costly development and maintenance for its mission-critical systems.  In addition to increased cost, any decrease in quality or increase in schedule could threaten DOD's warfighting adaptability and readiness.

DOD remains the key customer for Ada technology.  Although Ada 83 is being used outside DOD for the development of critical applications, the Ada tool and compiler market remains dependent on robust support from DOD.  Even the perception of DOD pulling away from its support for Ada could dramatically affect Ada vendors, at a time when the industry is in the process of assimilating Ada 95. Uncertainty over DOD's programming language policy and investment strategy is already affecting the ability to find capital to invest in Ada-related development.

The most critical impact of not sustaining Ada is the consequent reduction in support for DOD's 50 million existing lines of Ada mission-critical software.  Without DOD support, Ada will begin to resemble other unsupported DOD languages such as Jovial and CMS-2.  Mission-critical programs relying on Ada code will be forced to choose between spending time and money to keep their Ada support current and spending even greater resources to convert their software to another language.

## SUMMARY OF ADA TRENDS

Tables 1.1 and 1.2 summarize the differences between the past context (1970s and early 1980s) in which the current Ada policy was developed, and the current environment.  The change in context is sufficient to warrant a restructuring of DOD's policy and strategy for Ada.

Table 1.1  Past and Present Contexts for Ada:  General

| Past | Present |
|------|---------|
| Some chance for Ada to be a leading commercial language | Virtually no chance for Ada to achieve a commercial lead, except in niche areas |
| Some chance that Ada could drive other software practices | Virtually no chance for Ada to become a driver of other software practices |
| Fair chance that Ada could become the leading high-assurance, real-time (HA/RT) language | Ada generally considered the strongest (HA/RT) language in this area, but others widely used |
| New software mostly custom, requirements-driven | New software mostly (non-Ada) COTS-driven |

Table 1.2  Past and Present Contexts for Ada:  Department of Defense

| Past | Present |
|------|---------|
| DOD a dominant software player | DOD a large software player |
| Secondary role in DOD for software | Primary role:  key to DOD goal of information dominance |
| No DOD Ada legacy code | 50 million lines of DOD weapon systems Ada legacy code |
| DOD committed to major Ada development investment | DOD preparing to drop its investment in sustaining Ada |

## CRITICAL QUESTIONS

The discussion above indicates that there are serious problems with DOD's current policy regarding Ada, including the lack of guidance provided to DOD personnel and contractors for use of Ada, uneven implementation of the waiver process, and unrealistic investment strategies.  In the course of responding to its charge to recommend ways for improving DOD software policies and strategies regarding Ada, the committee identified several critical questions; they are summarized below and addressed in the remainder of this report.

- *What is the relationship between programming languages and software engineering practices?*  As embodied in DOD Directive 3405.1 (DOD, 1987a), DOD's current policy for software development is a programming language policy.  However, the choice of Ada as the programming language is insufficient to ensure the development of high-quality, reliable software systems for defense missions.  Chapter 2 addresses the importance of software engineering practices and their relationship to programming languages, and points out connections that DOD policy should take into account; Chapter 4 discusses implementation of a broader DOD software policy.
- *Are there application areas where using Ada makes an appreciable positive difference?*  In application areas where powerful non-Ada commercial software support is available, Ada is unlikely to be cost-effective.  However, for some DOD software applications, there are few commercial counterparts, and Ada may have advantages.  These issues are discussed in Chapters 2 and 3.

•   *If Ada is superior in some areas, is a policy requiring its use appropriate?*  This issue and a number of other policy alternatives are discussed in Chapter 3.

•   *For whatever policy requirements are appropriate, can DOD establish a workable set of criteria and processes for recognizing exceptions?*  These issues are discussed in Chapter 4 and are addressed in committee-suggested modifications of a May 1996 DOD draft software management policy presented in Appendix A.

•   *What specific investment strategies are needed to keep Ada strong?*  This issue is discussed in Chapter 5.

## NOTES

1. The cost to DOD of developing Ada 95 has been estimated by the Ada 95 program manager, Christine Anderson, as being in the range of $29 million to $35 million (personal communication, July 5, 1996).  Ada 83 investments were likely much greater but are difficult to quantify.

2. For descriptions of non-DOD projects using Ada, particularly in aerospace, transportation, and telecommunications, see "Ada Success Stories," maintained by the Ada Information Clearinghouse, located on the World Wide Web at http://sw-eng.falls-church.va.us/AdaIC/usage.

3. If all programming activities are included (i.e., application-oriented programming by other professionals), the total number of programmers increases to 3.45 million, less than 3 percent of which are Ada 83 programmers.

4. The newspapers were the April 21, 1996, edition of the *Los Angeles Times* and the March 24, 1996, edition of the *Washington Post*.

5. Other significant 3GLs were Cobol (11 percent) and Java (4 percent).  Significant 4GLs were Visual Basic (13 percent), PowerBuilder (10 percent), and FoxPro and Visual C++ (4 percent each).

6. The industry groups surveyed were automobile services, financial services, medical devices, and industrial machinery.

7. Exceptions to this generalization include Cobol, which was never a popular language in academia but is used widely in business and defense applications, and Pascal, which was popular for many years in universities but not in industry.

8. "Foundation" is defined as one of the initial computing courses taken by students majoring in the field.

9. Bob Mathis, executive director, Ada Resource Association, personal communication, September 8, 1996.

10. The committee included representatives of two firms that sell Ada products: Rational and Intermetrics.

11. In a position paper to the committee, Victor Vyssotsky advocated stronger encouragement for programmers to learn numerous languages (Vyssotsky, 1996).

12. As detailed above, only one-third of the 3GL code written for weapon systems is in Ada.  Because any software produced since the policy went into effect in 1987 (except for software not maintained or upgraded by DOD) would require a waiver, many more than a few waivers each year should have been approved for weapon systems alone.  In addition, the use of Ada for automated information systems in DOD is even lower in relative and absolute terms.

13. Gerald Pasternack, Citibank, presentation to the committee, May 23, 1996, Washington, D.C.

# 2

# Software Engineering and the Role of Ada in DOD Systems

Recent progress in software engineering includes the development of models and technology to improve software processes and architectures. This chapter highlights some of these approaches as a framework for crafting a DOD software policy that is more broadly conceived than the current policy on programming languages, and as a method of evaluating Ada's role in DOD systems. In addition to evaluating Ada's capability for supporting software engineering processes, this chapter compares Ada with other third-generation programming languages (3GLs). Technical comparisons between Ada and other languages can be made with greater confidence than can quantifying the performance of programming languages. The committee found the quality of the data available for such empirical analyses to be lacking; in addition to discussing these data limitations the committee thus suggests ways to improve collection of and access to needed data. In the absence of data that are reliable enough to serve as a basis for sound conclusions, the committee's findings are based on a combination of those data, technical comparisons, anecdotal project experience, and, ultimately, on the deliberations of the committee and its expert judgment.

## SOFTWARE ENGINEERING PROCESS AND ARCHITECTURE

In the 1980s, when DOD's current programming language policy was first established, implementation of such a policy was perceived to be the most straightforward approach to improving DOD's software engineering capability. But in the past decade software engineering technology and practices have changed fundamentally. Examples of important developments include advanced tools and techniques such as computer-automated software engineering (CASE) tools, application generators, and object-oriented methods; process improvement, including iterative/spiral development processes, the Software Engineering Institute's Capability Maturity Model, the Air Force Software Development Capability Evaluation, and the ISO 9000 quality standard; product-line management such as architecture-driven processes and components and common operating environments; and technology for heterogeneous software, including Open Systems, Internet, and Common Object Request Broker Architecture standards.

Reflecting these fundamental changes, a consistent theme emerged throughout the committee's deliberations and in presentations from industry and DOD experts: programming language is important, but not as important as a thorough understanding of application requirements, a mature software-development process, and a good software architecture.  While understanding of requirements is certainly an important factor in project success, it represents a largely language-independent aspect of software engineering and thus is not emphasized in the following discussion.  Process maturity and architecture quality, on the other hand, represent aspects of software engineering that are influenced and supported by programming languages and the environments in which they are developed.  Furthermore, one very important aspect of "good" processes and architectures is their ability to accommodate changes in requirements.[1]

While there is much debate over what constitutes an architecture,[2] the following observations are made to summarize architecture's importance and its close linkage with modern software development processes:[3]

• Achieving a stable software architecture represents a significant project milestone at which the critical decisions to make, buy, or reuse software should have been resolved.  This  milestone in a project's life cycle represents a transition from the exploratory stage (characterized by discovery and resolution of numerous unknowns and sources of risk) of a project to the development phase (characterized by management according to a particular development plan).[4]

• Architecture representations provide a basis for balancing the trade-offs between the problem space (i.e., requirements and constraints) and the solution space (i.e., the product design, implementation, and quality).

• The software architecture and process provide the framework for most of the important (i.e., high-payoff or high-risk) human communications among the analysis, design, implementation, and test activities.

• Poor architectures and immature processes often underlie many project failures.  A mature process, an understanding of the primary requirements, and a demonstrable architecture are important prerequisites for predictable outcomes.

• Architecture development and process definition are the intellectual steps that map a problem to its solution without violating existing constraints; these tasks require human innovation and cannot be automated.

DOD's formulation of an improved policy regarding its use of the Ada programming language should take into consideration the fundamental need for improved software architectures, more effective and mature development processes, and increased process automation.  Because DOD's requirements for quality—generally high reliability, state-of-the-art performance, and maintainability by DOD personnel—usually cannot be compromised, DOD software development projects often require increased funds and/or extended schedules.  The four subsections below describe process and architecture as elements fundamental to needed improvements in software economics—it is this significance of architecture and process that motivates the committee's belief that DOD should expand its software policy to encompass more than just a programming language policy.  The discussion below focuses on improving the cost-effectiveness of DOD software and achieving a better return on investment (ROI); it is assumed that quality is held fixed at the levels necessary for DOD systems.

## Economics of Software Engineering

Most software development costs are a function of four basic parameters:

1. The *size* of the software end product in terms of human-generated source code and documentation;
2. The *process* employed to produce the software;
3. The *environment* and tools employed to produce the software; and
4. The expertise of the *personnel* involved, that is, the capability of the software developers.

One very important aspect of software economics (as represented in today's software cost models) is that the relationship between cost and size exhibits diseconomies of scale: the cost per unit of functionality increases with software size. This relationship stems primarily from the complexity of managing interpersonal and interteam communications; as the number of team members increases, the complications relating to the team members' differing perspectives and backgrounds increase even more rapidly. Given the factors affecting software development costs, economic leverage is best achieved by focusing on technologies that enable the following:[5]

1. Reducing the size or complexity (or improving the architecture) of what needs to be developed;
2. Improving the development process; and
3. Employing better environments and tools to automate the process.

Most software experts would distinguish among the above factors; they would also acknowledge significant interrelationships. For example, tools enable reduction of the amount of source code and process improvements; attempts to reduce size lead to process improvements; and improved processes drive tool requirements. These interrelationships mean that even though programming languages do not directly affect project outcomes, they can have significant indirect effects.

## Reducing the Complexity of Software Products

In general, the most significant step toward reducing cost and improving ROI is to design an architecture that achieves product requirements and quality goals with the minimum amount of human-generated source material. This is the primary motivation behind development of high-order languages (e.g., Ada 83 and 95, C++, and fourth-generation programming languages), use of automatic code generators (CASE tools and graphical user interface builders), reuse of commercial off-the-shelf (COTS) software products (operating systems, "windowing" environments, database management systems, "middleware," and networks), and reliance on object orientation (encapsulation, abstraction, component reuse, and architecture frameworks).

Since the difference between large and small projects has a greater than linear impact on life-cycle cost, using the highest-level language and appropriate COTS or non-developmental items can lower costs significantly, especially in warfighting domains where large-scale systems are the norm. Furthermore, simpler is generally better: reducing its size usually makes a program more understandable, easier to change, and more reliable. One typical negative side effect is that higher-level abstractions tend to degrade performance—i.e., increase resource consumption, whether in processor cycles, memory usage, or communications bandwidth. Fortunately, these drawbacks have been greatly offset by improvements in hardware performance, compiler technology, and code optimization (although much less so in embedded platforms). Ada, and particularly Ada 95, allows for reduction in the source size of software products through language features that support abstraction, object-oriented programming, and component integration (e.g., of reusable components, COTS products, and legacy components). The language C++ provides similar advantages in the commercial market.

In numerous DOD domains, a common approach today is to maximize the use of COTS products.[6] While this is certainly desirable as a means of reducing the overall amount of custom development, it has often not been a "silver bullet" in practice. Table 2.1 identifies some of the advantages and disadvantages of employing COTS products, compared to custom software, in DOD domains.

The advantages of using COTS rather than custom software are significant, but there are still application areas (particularly in warfighting) in which the advantages of having control over reliability, performance, or rapid enhancements provided by custom software are compelling. The disadvantages of COTS are not sufficient to peremptorily drop that approach, but they point to areas in which architectural trade-off analysis and risk management approaches will be needed.

## Improving Software Processes

The importance of a mature software development process has been well established (CSTB, 1987; DOD, 1987b). Modern software development processes have moved away from the conventional waterfall model, in which each stage of the development process depends on completion of the previous stage. While there are variations, current concepts call for an initial version of a system to be constructed rapidly early in the development process, with an emphasis on addressing the high-risk areas, stabilizing the basic architecture, and refining the driving requirements (with extensive user input where possible). Development then proceeds as a series of iterations ("spirals," "increments," "generations," "releases," and other terms have been used), building on the core architecture until the desired level of functionality, performance, and robustness is achieved.

Software cost models, such as the COCOMO model (Boehm et al., 1996), have been updated to reflect the use of modern software development processes and can be used to quantify the importance of process. The parameters defining the effects of process on the cost and schedule estimates produced by COCOMO include the following:

- *Application familiarity*—the developer's degree of domain experience;
- *Process flexibility*—the degree of contractual rigor, ceremony, and freedom of change inherent in the project "contract," "vision," and "plan";
- *Architecture definition and risk resolution*—the degree of technical feasibility demonstrated prior to commitment to full-scale production;
- *Teamwork*—the degree of cooperation and shared vision among stakeholders (buyers, developers, users, and personnel responsible for verification, validation, and maintenance, among others); and
- *Software process maturity*—the maturity level of the development organization, as defined by the Capability Maturity Model (Paulk et al., 1993).

Cost estimates produced by COCOMO 2.0 show that the difference between a good and bad process for a large (300,000 lines of source code) program will often exceed a factor of 1.3 in the length of time it will take for a team to develop a software product, a factor of 2 in cost, and a factor of 5 in quality (delivered defect rate). Realization of this relationship has led to significant investments and advances in software process improvement techniques over the past 10 years, exemplified by DOD investment in the Capability Maturity Model, developed by the Software Engineering Institute at Carnegie Mellon University.

Table 2.1  Advantages and Disadvantages of Commercial off-the-shelf (COTS) and Custom Software

| Integration of COTS | Custom Development |
|---|---|
| **Advantages** | |
| • Predictable license costs | • Complete freedom |
| • Broadly used, mature technologies | • Smaller, often simpler |
| • Immediately available | • Often better performance |
| • Dedicated support organization | • Control of development and enhancement |
| • Hardware/software independence | • Control of reliability trade-offs |
| • Rich functionality | |
| • Frequent upgrades | |
| **Disadvantages** | |
| • Up-front license fees | • Development expensive, unpredictable |
| • Recurring maintenance fees | • Availability date unpredictable |
| • Dependence on vendor | • Maintenance expensive |
| • Sacrifices in efficiency | • Portability often expensive |
| • Constraints on functionality | • Drains expert resources |
| • Integration not always trivial | |
| • No control over upgrades and maintenance | |
| • Unnecessary features that consume extra resources | |
| • Reliability often unknown or inadequate | |
| • Scale difficult to change | |
| • Incompatibilities among vendors | |
| • Licensing and intellectual property issues | |
| • Difficulties in testing and evaluation | |

## Influence of Software Environments, Tools, and Languages on the Software Engineering Process

The tools and environment employed in the software engineering process generally have a linear effect on the productivity of the process. Compilers, editors, "debuggers," configuration management tools, traceability tools, quality assurance analysis tools, test tools, and user interfaces provide the foundation for carrying out the process and maximizing automation. However, the maturity, availability, and performance of these tools—and their procurement and maintenance costs—must be taken into account. Cost models indicate that tools and automation generally enable cost savings ranging from 30 to 60 percent (see Box 3.2 in Chapter 3).

As mentioned above, environments and tools have indirect effects; however, they also enable certain improvements in process and reductions in size that have much greater impacts. Thus, the view that the quality of the software engineering process is independent of the programming language can be misleading. Language standardization has led to tools for automated support of configuration control and increased automation of quality assessment (through interface specification, compilation and consistency analysis, readability, and "inspection automation"). These, in turn, have led to practical and significant process improvements, such as iterative development, architecture-driven design, and automation of documentation (Royce, 1990). Furthermore, languages like Ada 95 add object-oriented features, which have enhanced their versatility. Such features, in some cases, allow Ada 95 programs to implement the same function as Ada 83 programs with a significant reduction in the number of source lines of code.[7] These improvements are not unique to Ada 95, but absent the technical features and automation support that are standard in the Ada environment (compiler, library manager, and debugger), many of these process improvements have been impractical for other languages.

Over the past 10 years, Ada 83 has supported the process and software engineering design goals described above by enabling (1) integration of components (abstraction and encapsulation, language standardization, separation of module specification from body, strong typing, and library management) that allow structured design early in the life cycle and incremental improvement of the breadth, depth, and performance of the evolving design through multiple iterations; (2) reduction of rework via early definition and verification of architectural interfaces prior to coding; (3) incorporation of configuration-management discipline, separate compilation, and interface and implementation partitioning directly in the language, thus enabling environments that are much more controlled and are instrumented for single and multiple team development and management of continuous change; and (4) reliability features, allowing errors to be automatically identified earlier in the life cycle by compile-time and run-time consistency checks. Ada 95 further improves this language support.

The primary point of this section—that the software engineering community has benefited greatly from the use of Ada, owing mostly to the language's support for the transition to development of better processes and better architectures—was predicted by Fred Brooks a decade ago (Brooks, 1986):

> I predict that a decade from now, when the effectiveness of Ada is assessed, it will be seen to have made a substantial difference, but not because of any particular language feature, nor indeed of all of them combined. Neither will the new Ada environments prove to be the cause of the improvements. Ada's greatest contribution will be that switching to it occasioned training programmers in modern software design techniques.

It is from such a perspective that the committee analyzed the business case for use of Ada (Chapter 3) and as a result recommends a broader software engineering policy for DOD (Chapter 4).

## TECHNICAL EVALUATION OF ADA 95 AND OTHER THIRD-GENERATION PROGRAMMING LANGUAGES

This section provides a brief technical evaluation of the programming languages Ada 95, C, C++, and Java, based on the summaries of language features given in Appendix B and focusing specifically on attributes related to the development of real-time critical systems. The committee's evaluation led it to conclude that, for real-time critical systems, Ada 95 is superior to the other languages, from a technical and software engineering standpoint. It is important to recognize that some facets of this technical evaluation may change over the next several years as the other languages, particularly Java, mature and evolve in response to applications with requirements for higher integrity or real-time multimedia interaction, for example.

Criteria related to critical systems development fall into two sets of categories: (1) compile-time and run-time checking to support encapsulation and safety, and (2) support for hard real-time systems.[8] Criteria related to encapsulation and safety include:

- Support of user-defined abstractions and enforcement of modularity and information hiding;
- Compile-time enforcement of type distinctions;
- Run-time management of pointers, arrays, and variant structures; and
- Support for software fault tolerance and recovery.

With respect to these criteria, Ada 95 and Java fare well. Ada 95 offers more support than does Java for compile-time type distinctions by (1) employing generic templates, and (2) simplifying the expression of strong type distinctions between otherwise structurally equivalent numeric, enumeration, array, and pointer types. Java and Ada provide stronger enforcement of modularity and information hiding than do C and C++, because C and C++ both provide "back doors" that allow external access to internal variables. Java and Ada also provide the following safety features: (1) default "null" initialization of pointers, and run-time checks for null on all dereferences of pointers; (2) run-time checks for out-of-bounds indexing into an array and attempts to select the wrong variant from a subtype hierarchy; and (3) run-time exceptions indicating all failures of run-time checks, allowing programmer-specified exception handlers to implement appropriate fault tolerance and recovery actions. In C and C++, there are no checks to prevent the creation of dangling references to data structures, making the use of pointers more error prone.

Criteria related to real-time systems development include:

- Support for safe, static allocation of all run-time data structures;
- Predictability of constructs with respect to real-time deadlines; and
- Language support for real-time-oriented interactions between multiple threads of control.

With respect to these criteria, Ada 95 provides a number of advantages, including providing mechanisms for statically preallocating data structures while still allowing safe and convenient manipulation of such structures with pointers. In Java, all non-primitive data structures are allocated dynamically (on the "heap"), with the attendant danger of run-time storage exhaustion and unpredictable storage allocation and reclamation times. Using only "static" structures in Java could make this allocation predictable, but in many cases this restriction would create additional problems.

Ada 95 provides data-oriented synchronization mechanisms that reduce overhead and minimize the potential for high-priority threads being delayed indefinitely while waiting for release of resources held by lower-priority threads (priority "inversion"). Java provides some multithreading primitives in the language and the standard library, but the standard Java locking mechanism provides no standard support for limiting the amount of time a high-priority thread will wait for a low-priority thread. The interthread interaction model in Java is based on explicit notification rather than state-oriented guards, increasing the likelihood of race conditions, which can lead to uncertainty in data access. The C and C++ languages do not directly address multithreading and support for real-time processing.

For critical real-time code, Ada 95 emerges as technically superior compared to Java, C, and C++. The Java language has not yet been standardized and its design is still somewhat in flux, and it may evolve to provide further support for critical and/or real-time systems. C, C++, and Ada can also be expected to continue to evolve, albeit at a slower pace.

From a business case standpoint it is too early for DOD to consider Java in this application domain. Java might evolve into a language with strong real-time support capabilities, or it might not. For the foreseeable future, Ada provides the strongest available support for high-assurance, real-time software development. As languages develop attractive new capabilities, DOD should be prepared to periodically perform technical comparisons, such as the one provided here and in Appendix B. But as discussed in Chapter 3, such a technical comparison is only one part of the business case associated with establishing a software management policy.

## AVAILABLE COMPARISONS OF ADA 83 AND OTHER THIRD-GENERATION PROGRAMMING LANGUAGES

Over the past 5 years several studies have concluded that, for custom software development, Ada 83 is more effective than its leading alternatives (Cobol, C, C++, Fortran, Jovial, and CMS2) in improving software maintainability and reliability, improving overall life-cycle cost, and enabling management of the risks of large-scale development (Mosemann, 1991; Masters, 1996).  These studies are based on various mixes of expert opinion and project results.  The data supporting the conclusions are generally not normalized or controlled, and none of the studies to date has been rigorously peer reviewed.  This section summarizes the available information comparing Ada 83 with other leading 3GLs.  This information falls into three main categories:

1.  *Analyses of language features:*  comparisons of how the features of programming languages contribute to such desired properties as reliability, maintainability, and efficient programming;
2.  *Comparisons of empirical data:*  comparisons based on data collected from completed projects in various languages; and
3.  *Anecdotal experience from projects:*  qualitative responses to project outcomes.

### Analyses of Language Features[9]

Analyses of language features have the advantage of being based on full and open examination of well-defined language features.  Their main disadvantage is that they are partial at best and are particularly weak in assessing the complex trade-offs among language features accomplished in the course of actual projects.

The most thorough of these analyses can be found in a 1985 Federal Aviation Administration study (IBM, 1985) comparing Ada 83 with C, Pascal, Jovial, and Fortran, and in a 1991 Software Engineering Institute study (Weiderman, 1991) covering Ada 83 and C++.  Both studies used the same evaluation scales, covering the desired properties of capability, efficiency, availability/reliability, maintainability/extensibility, life-cycle cost, and risk.  Figure 2.1 summarizes the results of these analyses, comparing Ada 83 and C (IBM, 1985) and Ada 83 and C++ (Weiderman, 1991) to the theoretical maximum score (higher numbers indicate better performance; the full definitions of criteria and numerical results are provided in Appendix D).  The differences in the ratings for Ada 83 in the 1985 and 1991 studies are probably good indicators of the variability attendant on evaluations of this nature.[10] Allowing for the range of variability, the most significant difference shown in Figure 2.1 is Ada 83's much stronger rating in the availability/reliability area, corroborating the results of this committee's comparative analysis in the preceding section ("Technical Evaluation of Ada 95 and Other Third-Generation Languages") and in Appendix B.

### Comparisons of Empirical Data

The major advantage of empirical project data is that the data represent the end results of projects and reflect the various features of each language.  The major disadvantage is that the varying conditions associated with disparate projects make it difficult to assess sources of variability caused by differing definitions, assumptions, and contexts.  Moreover, many of the results come from proprietary (and thus unavailable) data on project productivity and quality, such as those presented in Jones (1994) and Reifer (1996).[11]  The major source of empirical data and information derived from them are summarized in Table 2.2  More details on the data are provided in Appendix D.

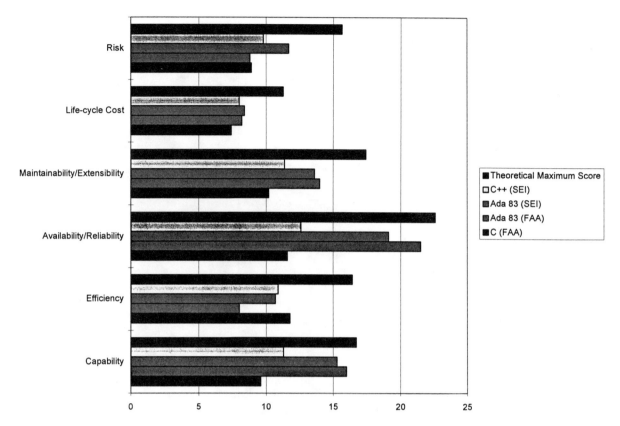

FIGURE 2.1  Comparisons of language features.  SOURCES:  Software Engineering Institute (SEI) data from Weiderman (1991);  Federal Aviation Administration (FAA) data from IBM (1985).

One major problem with empirical data is that quantitative determination of important features such as cost per source line of code (cost/SLOC) and defects per 1,000 source lines of code (defects/KSLOC) is confounded by differences in the expressive power of a source line of code in different programming languages.  One way of normalizing is to look at source lines of code per function point (Jones, 1995).  However, as shown in Table 2.3, these ratios have wide variability.  Lubashevsky (1996) reports variations in source lines of code per function point exceeding factors of 2 for C and 6 for C++.  Finally, there are differences in expressiveness for the same language across different application domains.

Appendix D points out that the similarity in the data for C, Ada, and C++ in Jones (1994) relating to cost per function point and defects per function point (see Table 2.2) appears simply to reflect Jones's (1995) mean values of SLOC/function point for C, Ada, and C++.  Thus, the Jones data appear to indicate that cost/SLOC and defects/KSLOC show little variation across programming languages.  While this conclusion is perhaps warranted for cost, it is a conclusion at considerable variance with the data from other studies on defects/KSLOC.

The different cost/SLOC values given by Reifer (1996) appear to be overshadowed by potential differences in the relative expressive power of a line of code in Ada, C, and C++.  However, the lower number of defects/KSLOC reported for Ada in Reifer (1996) is still significant, particularly with respect to embedded weapon systems software.

Table 2.2 Comparisons of Languages: Summary of Empirical Data

| (Study) Domain | Data, Analysis Available? | Number of Ada Projects | Number of Total Projects | Cost ($ per source line of code) | | | | Defects per KSLOC[a] | | | |
|---|---|---|---|---|---|---|---|---|---|---|---|
| | | | | For-tran | Ada 83 | C | C++ | For-tran | Ada 83 | C | C++ |
| (Zeigler, 1995) Compilers, tools | Yes | 1 | 2 | n/a | 6.6 | 10.5 | n/a | n/a | 0.10 | 0.68 | n/a |
| (NASA-SEL, 1994, 1995)[b] Engineering | Yes | 15 | 120 | Ada 83 40% lower than Fortran | | n/a | n/a | 4.8 | 2.10 | n/a | n/a |
| (Reifer, 1996) Military information systems | No | ? | 190 | n/a | 30 | 25 | 25 | n/a | 3.00 | 6.00 | 4.00 |
| (Reifer, 1996) Embedded weapon systems | No | ? | 190 | n/a | 150 | 175 | n/a | n/a | 0.30 | 0.80 | 0.60 |
| (Jones, 1994)[c] Telecommunications | No | ? | Many | n/a | 1760 | 2966 | 1180 | n/a | 0.17 | 0.29 | 0.14 |

[a] KSLOC: 1,000 source lines of code.
[b] McGarry et al. (1994); Waligora et al. (1995).
[c] Values in terms of function points, rather than lines of code.

Table 2.3  Source Lines of Code per Function Point

| Language | Low | Mean | High |
|----------|-----|------|------|
| Ada 83 | 60 | 71 | 80 |
| C | 60 | 128 | 170 |
| C++ | 30 | 53 | 125 |

SOURCE:  Data from Jones (1995).


Two sources of data are particularly sound with respect to comparability of projects and availability of data and analysis.  The Zeigler (1995) study has the most thorough analysis of whether the differences in data on the cost and number of defects for Ada and C could be caused by other factors. With respect to expressiveness of a line of code, Zeigler analyzed lines of code per feature (LOC/feature) and found that Ada was about 16 percent more "verbose" than C (109 LOC/feature for Ada, compared to 94 for C).  Applying this correction to the data indicates that Ada outperforms C on the basis of cost per feature by a factor of 1.37 and on the basis of defects per feature by a factor of 5.9.

Zeigler also analyzed the potential for confounding effects of relative software complexity, personnel capability, and learning curve effects associated with C and Ada, and presents a good case that these factors did not cause any significant bias.  Zeigler thus provides a strong argument for Ada programs having had lower life-cycle costs and fewer defects than C programs in the large (more than 1 million lines of both Ada and C) project in compilers and tools that was the basis for his study. However, the Zeigler study is suggestive, rather than definitive, about the applicability of this result to other domains and other teams (the development teams on the projects studied were stable and composed of highly capable, seasoned personnel).

The NASA Software Engineering Laboratory (SEL) projects described by McGarry et al. (1994) also provide some helpful comparative data but do not cover most DOD domains of interest.  The SEL projects are highly precedented, flight dynamics engineering applications that are not embedded and do not require real-time functionality.  SEL's analyses of Ada initially concluded that, owing to greater reuse, Ada projects enjoyed significant cost and schedule reductions compared to those using Fortran. Subsequently, application of the Ada object-oriented reuse approach in Fortran projects yielded comparable gains.  Significantly, however, over the period from 1988 to 1994, the rate of defects associated with Ada was less than half the level of defects seen with Fortran.

A subsequent NASA-SEL study (Waligora et al., 1995) corroborated the reduction in defect rate with Ada, and concluded that Ada development costs were 40 percent less than those of Fortran, for equivalent functionality.  This conclusion was based on analysis indicating that Ada's generic features achieved reuse with many fewer statements than Fortran's repeated code.

Both the SEL and Zeigler analyses also concluded that programming languages were not the dominant factor in influencing software productivity and quality.  SEL found several other variables (object-oriented reuse, use of Cleanroom techniques, code reading) to be more significant.  Zeigler cites architecture and design, configuration management, testing, process, programmer expertise, and management skills as more significant than the particular programming language used.

In summary, based on analysis of available empirical data and comparisons of language features, a conclusion that Ada is superior in ensuring availability, reliability, and fewer defects appears warranted.  The evidence is not strong enough to assert Ada's superiority with respect to cost, but when considered with other data (Appendix D), and given the lack of solid evidence indicating less expensive custom software development in other languages, a case can be made that using Ada provides cost savings in building custom software, particularly for real-time, high-assurance warfighting applications.

**Anecdotal Experience from Projects**

The current DOD base of experience with Ada is substantial.  In DOD software inventories, Ada represents about one-third of weapon systems code (Hook et al., 1995),  representing more than 50 million source lines of operational code in many of DOD's most crucial systems.

The Aerospace Industries Association (AIA) has stated that its member companies have all ". . . greatly benefited from the software engineering support features of Ada, including reduced error rates," and notes that Ada has had a "substantial positive impact" especially in "large, high visibility projects such as F22, BSY-2, Boeing 777, and Peace Shield" (AIA, 1996).  However, AIA also notes that "some [projects] have suffered because Ada support tools were not robust nor available when needed, or because Ada presented interface difficulties in heterogeneous environments" (AIA, 1996);  AIA has advocated ending DOD's Ada mandate.

Representatives from the DOD services related to the committee numerous instances of success in Ada projects that were delivered (to varying degrees) on budget, on schedule, and with satisfied users. Some of the most compelling data in this regard, drawn from a broad range of projects, were provided by Robert Kent of the Air Force Electronic Systems Center (ESC).  According to Mr. Kent, ESC's experience is that "Ada projects have a much higher success rate than non-Ada projects."  He substantiated this claim with several case studies indicating that a substantial number of very large mission-critical applications (greater than 1 million source lines) have been successfully delivered and maintained in Ada.  While the financial success of Ada projects is not universal, some of the results from the case studies indicate that a mature software organization will perform better with Ada than with other languages.

Other case studies presented to the committee to illustrate successful Ada development include the following:

- Air Force:  CCPDS-R, Cobra Dane System Modernization, REACT, STARS demonstration project,[12] PRISM;
- Navy:  BSY-2, AEGIS; and
- Army:  FAADC2I.

Ten years ago, it was very difficult to find a single software success story in any programming language.  Today, there are several, and most of the large-scale, successful DOD projects have employed Ada as one of the technologies to support their efforts to improve both processes and architectures.

Successful Ada users outside DOD include DOE (AdaSAGE), NASA (Space Station program and Software Engineering Laboratory), numerous international organizations (Transport Canada, Canadian Department of National Defense, Celsius Tech, Eurocontrol, Australian Commonwealth, United Kingdom Ministry of Defense).  Commercial organizations have also utilized Ada in their products.  A primary example is Boeing Corporation, which, like DOD, sought a single, common programming language for its commercial mission-critical software (Box 2.1).

**THE NEED TO INSTITUTE COLLECTION OF DATA FOR SOFTWARE METRICS**

The committee searched for sources of data that could provide a strong scientific basis for concluding that Ada is or is not a superior programming language in any given application domain.  With respect to such confirmatory data, the committee concluded the following:

## BOX 2.1
## Use of Ada on Boeing Commercial Airplanes

*Decision to use Ada.* In the late 1970s, Boeing began to use airborne software on its 757 and 767 airplanes. Due to the state of practice, a large variety of languages and language processors were used, thus making the application of standards difficult. In 1985, Boeing's Commercial Airplane Group (BCAG) initiated a program to solve the airborne software language problem. The first step in the initiative was to choose a preferred language. The major elements of the criteria were support of structured programming practices, high-order language, block structured language, portability, and understandability. Additional goals included the use of software engineering principles such as information hiding, abstract data types, and strong type checking; the ability to specify interfaces precisely; and the use of standardized fixed-point and floating-point arithmetic. The evaluation process involved consulting with several key suppliers. The process resulted in the selection of the Ada language.

*Preparation to use Ada.* BCAG relies on its suppliers to provide airborne software for its commercial airplanes. Hence, preparation to use Ada had to be a joint program with its suppliers. The charter of the joint program included evaluating compilers, preparing personnel to use Ada, and sharing of operational experiences in the use of Ada. Several joint meetings were held over a period of years, and a newsletter was published to share information and insights. Boeing also prepared guidelines to define a subset of Ada for use in safety-critical applications. The guidelines benefited from input from the joint program and other industry sources.

*Experience with Ada.* The use of Ada has significantly reduced the number of different programming languages used on the Boeing 777. Ada was used on 60 percent of the systems on the 777 and represented 70 percent of the lines of code developed. No correlation was found between the language used and the number of problems found on a system. The other principal language used in new development was C. The richness and complexity of the Ada language helped knowledgeable users with mature tools achieve modest gains in productivity. However, the complexity of the language caused problems for other users who had to work through compiler problems. A key lesson learned was that the need for retraining was not adequately understood.

*Future plans for use of Ada.* BCAG expects to continue its use of Ada for airborne applications. A standard language allows the use of tools to aid in the development of software that would be difficult or impossible to implement in a multiple-language environment. The use of Ada in the future would be improved by greater consistency among the available compilers.

SOURCE: Leonard L. Tripp, Boeing Commercial Airplane Group, personal communication, August 27, 1996.

- The data are *uneven.* The Ada community has collected a good deal of data on DOD's and other organizations' experience with Ada, but comparable data are not available on DOD's experience with C, C++, and other languages.
- The data are largely *unavailable.* The software data that have been collected are largely in proprietary databases held by DOD contractors, consultants, or commercial cost-modeling or market-analysis firms.

- The data *lack common points of comparison and are incommensurable*. Some of the data are known to be inconsistent with respect to rules for counting lines of code, functionality, effort, and defects. Other data are accumulated with no knowledge of their degree of consistency.
- The data are *incomplete*. Most of the data collected address quantities such as size, level of effort, and number of defects, but do not take into account the environmental variables (e.g., cost drivers) associated with the quantities measured. Also, it is unclear to what extent the data collected are fully representative of project experience (e.g., perhaps only the good projects collect or report data).
- Data are *collected but are not systematically organized*. Many DOD organizations collect software data for project monitoring and control, as well as environmental data such as those indicating process maturity. But the data are not organized and stored in a repository that could facilitate analysis to support DOD software engineering process- and product-improvement efforts and policy analyses.

Most of the available data generally support the conclusion that Ada is preferable for DOD warfighting applications, and the committee did not find any data to refute that conclusion. But for future policy analyses and initiatives to improve DOD software engineering practices, a stronger base of DOD software metrics data that describes project outcomes would more than repay the investment necessary to develop it. Without more reliable data, decision making will have only a weak foundation.

On an individual project level, DOD has endorsed the concept of using metrics data to improve software process management through its endorsement of the Software Engineering Institute's Capability Maturity Model, which includes quantitative process management as a key process. Within DOD, several local efforts to collect and analyze data for evaluating software add considerable value. Some good examples are the Air Force Cost Analysis Agency, the Army Software Test and Evaluation Program, the Navy Undersea Warfare Center, and the Defense Logistics Agency's Columbus, Ohio, facility. However, DOD is not applying this practice at a more strategic level to improve its overall software cost-effectiveness and is missing a major opportunity to improve its software cost-effectiveness. DOD should establish a sustained commitment to collect and analyze consistent software experience data.

Foundations for such a program already exist. The Joint Logistics Commanders "Practical Software Measurement" guidebook (DOD, 1996a) provides good case studies and guidelines for tailoring and focusing measurement of software capabilities on activities that add value. The Software Engineering Institute's software core metrics reports (Carleton et al., 1992) provide a foundation for collecting consistent data across projects and organizations. NASA's Software Engineering Laboratory (McGarry et al., 1994) provides a good model. Also, a good start toward a DOD software metrics initiative is represented by its National Software Data and Information Repository; this effort needs some improvement and has languished due to lack of a sustained commitment.

As DOD's chief information officer, the Assistant Secretary of Defense ($C^3I$) is the logical focal point for establishing and sustaining a DOD-wide software metrics initiative. The initiative would need a precise scope, strong staffing, and focused management, but the examples above provide evidence that such investments can generate significant positive results.

## NOTES

1. Barry M. Horowitz, president of MITRE Corporation, has noted the following with respect to requirements and architecture: "Both government and industry typically put almost all of their efforts into the initial performance and functionality of a program in spite of the fact that these will change substantially over the life of the system. At the same time, there is a near-total lack of attention to an architectural baseline that would form a stable foundation for incorporating the system's changing requirements" (Horowitz, 1991, p. 10).

2. No technical standard exists for software architecture; however, the IEEE Software Engineering Standards Committee has created a planning group to investigate the issue. See "Standards Annual Report—1996," located at http://www.computer.org/standard/anreport/toc.htm.

3. Horowitz (1991) emphasizes the importance of architecture; more recently, a Defense Science Board Task Force emphasized the importance of software architecture, and estimated that "a well-formulated architecture might reduce costs of changes/upgrades by 30-50%" (DOD, 1994a).

4. See the contents of the life-cycle architecture milestone and associated rationale in Boehm (1996).

5. While personnel capability and understanding of requirements are also important cost factors, these topics are excluded from the discussion below because they are mostly language-independent variables.

6. This trend is driven by the dual-use initiative within DOD (DOD, 1995b) and by legislative changes, namely the Federal Acquisition Reform Act of 1996.

7. One source (Jones, 1995) found that the mean value of source code statements per function point is 71 for Ada 83 and 49 for Ada 95, for a 30 percent reduction. The amount of empirical evidence is small, however.

8. Other sets of criteria would apply for other classes of applications.

9. No independent evaluations of language features were located by the committee, prompting the analysis presented above in the section titled "Technical Evaluation of Ada 95 and Other Third-Generation Programming Languages" and in Appendix B of this report. Most evaluations have been carried out by government agencies or at their direction.

10. Some of the differences in efficiency and risk ratings may be due to increased Ada maturity, but the decline in availability/reliability is more likely due to differences in interpretation of the evaluation criteria.

11. Both of these authors are software consultants; Capers Jones is president of Software Productivity Research Inc., and Donald Reifer, formerly director of DOD's Ada Joint Program Office, is with Reifer Consultants Inc.

12. See Frazier and Bailey (1996) for a recent discussion of STARS demonstration project outcomes.

# 3

# DOD Software Policy: Analysis and Recommendations

The kinds of software programs that DOD needs for weapons systems and those that commercial industry creates for popular use often do not share the same basic requirements. DOD's requirements for high assurance—which include reliability, availability, survivability, safety, and security—are considerably more stringent, given the national security stakes involved. DOD's requirements for performance-critical real-time embedded operation continually stress the capabilities of available hardware and software technology. Its requirements for complex integrated systems mean that weapon systems software must perform at a level higher than typical commercial programs found at the same stage of development, and all of these requirements must be satisfied simultaneously. In this regard, DOD (and its primary contractors) continues to be at the leading edge of the development of large-scale software engineering technology. It is likely to remain in this position for many years to come, although an increasing number of commercial software systems are comparable to those of DOD in size and required robustness.

This chapter establishes a set of objectives and criteria for evaluating the cost-effectiveness of alternative programming languages for support of DOD missions. It then presents a business-case analysis to evaluate Ada against other programming languages (generally focusing on C and C++) with respect to these criteria. Drawing on the technical and empirical evaluations reviewed in Chapter 2, this analysis is performed for (1) DOD warfighting software with no direct commercial counterparts, and (2) commercially dominated applications that are commonly used in all organizations. The analysis concludes that Ada can provide DOD with a considerable advantage in DOD-dominated warfighting applications, but not in commercially dominated applications. It also concludes that ensuring the Ada advantage for warfighting applications will require DOD investment to sustain a robust Ada infrastructure, but that the benefits justify the investment.

The final sections of this chapter summarize the policy changes recommended by the committee on the basis of this analysis, evaluate this policy with respect to possible alternative DOD policies on Ada, and present an economic evaluation of the recommended investment strategy.

## POLICY OBJECTIVES AND CRITERIA RELEVANT TO MEETING THEM

### Relating Criteria to Objectives

DOD increasingly is emphasizing the concept of "information dominance" as a key to military superiority (Powell, 1992; DOD, 1996d). Secretary of Defense William Perry has stated that ". . . our warfighting strategy sustains and builds . . . on the application of information technology to gain great military leverage to continue to give us [an] unfair competitive advantage" (Perry, 1996a).[1] Since software development is a fundamental aspect of information technology, an appropriate objective for DOD software policy is to ensure that DOD software enables systems that are superior to any that an adversary might develop, thus affording DOD a competitive advantage.

Achieving superiority in software and information technology for military purposes requires attention to situation-dependent mixes of functionality and other attributes. In an electronic warfare countermeasure/counter-countermeasure competition, rapid and reliable modification of software is a critical capability. To thwart information system penetration, subversion, and threatened denial of service, computer security is paramount. Performance is critical for processing of wideband sensor data. Safety is critical to controlling the course of weapons. Interoperability is critical for joint operations. And for all DOD systems, DOD's budget limitations make cost critical.

It would be convenient if these criteria could be combined into a single "DOD return on investment" index. Alternative policies could then be evaluated definitively with respect to this composite criterion. However, decades of policy analysis have led to the conclusion that such combined analyses are feasible only in the context of particular military scenarios (Quade, 1964; Quade and Boucher, 1968). This creates two equally difficult problems—aggregating the results from multiple scenarios, and evaluating the relevance of the scenarios in a rapidly changing world.

Given this situation, a useful approach is to identify the critical criteria most relevant to DOD's achieving information dominance, to evaluate Ada and alternative programming languages with respect to these criteria, and to base conclusions and recommendations on a judgment-based weighting of the criteria. Given the objective of having superior software, such evaluation must consider the effect of DOD programming language choices both on DOD itself and on its adversaries.

### Critical Criteria in DOD's Selection of a Programming Language

As pointed out in "Economics of Software Engineering" in Chapter 2, software cost modeling has shown that the criteria of software functionality, cost, and speed of development can be related to other criteria that are more closely coupled to choice of programming language. These determinants are software size, process, development environment, and personnel. Additional criteria for assessing software quality that are crucial in supporting DOD's critical missions and ability to respond rapidly to changes include *high assurance and real-time performance,* a set that covers reliability, security, safety, survivability, and real-time performance; and *ease of change,* a capability that enables rapid response to changes in threats, technology, or mission requirements.

Besides the individual production factors involving software tools, technology, and personnel, an even stronger determinant of international competitive advantage in information dominance is the existence of a *socio-technical infrastructure,* which couples the production factors with knowledge resources, marketing channels, strategic partnerships, user groups, closely linked customer-supplier chains, and trends in domestic demand, thus stimulating innovation and enabling rapid development of new software products. Porter's (1990) analysis of the significance of a strong national socio-technical infrastructure in determining the competitive standing of an industry or a service sector supports consideration of the socio-technical infrastructure's role in the warfighting sector also.

These seven criteria—software size, process, environment, personnel, high assurance and real-time performance, ease of change, and socio-technical infrastructure—are used in the business-case analysis that follows.

## Warfighting and Commercially Dominated Applications

The current socio-technical infrastructure is probably the main factor considered by many commercial firms in choosing C or C++ as their primary programming language, even when they may have performed a technical evaluation that indicated Ada was superior.[2]  For example, the fact that much of the financial community's software expertise, tooling, components, and middleware is oriented to C and C++ would generally be enough to lead some firms to choose C or C++.[3]  Conversely, this argument can be turned in favor of DOD's maintaining an Ada-based competitive advantage in software for warfighting, a domain in which DOD of course plays the dominant role.

The term "warfighting" is used in the business-case analysis that follows to determine whether the use of Ada would provide a clear advantage for DOD, when compared to the alternative of using (typically non-Ada) commercial software solutions.  This determination should be made at the subsystem level, in the context of decisions to build or buy software for the system.  For example, an integrated combat system such as AEGIS can have "warfighting" weapon control subsystems for which a custom Ada solution is superior, and "non-warfighting" data management and graphical user interface subsystems for which non-Ada commercial solutions are superior.

### *Warfighting Applications*

There are two primary criteria for determining whether a subsystem belongs in the "warfighting" category:

1.  Relatively little commercial software and expertise is available for implementing the desired functions.  For example, even though intelligence analysis is involved in warfighting, many of its functions (database update, query and visualization, report generation) can be readily satisfied via non-warfighting commercial software.

2.  The application requires software quality attribute levels higher than those supportable by commercial software.  For many warfighting functions, these involve real-time performance, reliability, and survivability, particularly in high-stress, crisis-mode situations in which DOD information processing functions may be under attack.

The application domains for warfighting software include, but are not necessarily limited to, the following areas.  Also mentioned are related support services and capabilities that are nevertheless outside the very specialized domain of warfighting.

•  *Weapon control* includes weapon sensor processing; guidance, navigation, and control; and combat-oriented weapon-delivery platform control.  Included also are special weapon delivery platform operator devices such as heads-up displays.  Weapon control does not include administrative functions and "hotel services" for large weapon delivery platforms such as aircraft carriers, or support subsystems performing mainstream data management, networking, and graphical user interface functions.

- *Electronic warfare* depends on software involved in rapid-response electronic detection, identification, discrimination, tracking, platform-based communication, and associated countermeasure and counter-countermeasure applications. It does not include software for support subsystems performing mainstream data management, networking, and graphical user interface functions.

- *Wideband real-time surveillance* includes processing of hard or soft real-time images and data from infrared, radar, or other sensors. It does not include off-line query and analysis of surveillance archives, or support subsystems performing mainstream data management, networking, and graphical user interface functions.

- *Battle management and battlefield communication* includes hard or soft real-time weapons allocation, targeting, control, coordination, damage assessment, and associated battlefield communications requiring such special capabilities as spread spectrum, antijamming, and frequency hopping. It does not include off-line monitoring, update, query, and analysis of battle asset status, or off-battlefield communications. Thus, the range of "warfighting command, control, and communications ($C^3I$) applications" is narrower than previous categorizations such as "$C^3I$" or "mission critical."

In addition to the nonwarfighting, more generic applications mentioned above, warfighting also does not include testing, simulation, training, off-line analysis, maintenance, and diagnostics. Software for such applications might well be implemented in Ada, but other languages may be better choices in some situations.

### *Commercially Dominated Applications*

Commercially dominated applications include office and management support, routine operations support, asset status monitoring, logistics, medicine, and non-battlefield communications processing.

## ADA BUSINESS-CASE ANALYSIS

Table 3.1 summarizes the results of the committee's business-case analysis for DOD use of Ada versus other third-generation programming languages (3GLs) for both warfighting and commercially dominated applications. Evaluation of Ada with respect to software size, process, development environment, personnel, high assurance and real-time performance, ease of change, and socio-technical infrastructure is presented below.

### Criteria for Evaluation of Ada

#### *Software Size*

The critical portions of warfighting applications are largely custom software, or components reused from previous defense applications. Reducing the size of these applications involves capitalizing on the existing applications software base. The largest fraction of this software base (approximately one-third) is in Ada, giving an appreciable (but not overwhelming) advantage to Ada. Furthermore, dropping Ada would leave DOD with a large body (approximately 50 million source lines) of Ada code to re-engineer.

Table 3.1  Ada Business-Case Analysis Summary

| Criterion | Warfighting Applications | Commercially Dominated Applications |
| --- | --- | --- |
| Size | • Critical portions largely custom, or DOD-managed reuse<br>• Some Ada advantage due to existing DOD Ada investments<br>• Dropping Ada would leave large existing DOD Ada software base unsupported | • Predominantly commercial off-the-shelf (COTS)-based solutions; much less programming than with custom solutions<br>• COTS largely non-Ada, with many volatile interfaces; a considerable disadvantage for Ada |
| Process | • Some Ada advantage:  early verification of architecture interface obviates need for rework | • Ada advantage for custom software is less applicable for COTS; some Ada disadvantage in COTS-based rapid development processes |
| Environment | • DOD investment required for Ada parity in general tools<br>• DOD investment creates some Ada advantage in high-assurance, real-time tools | • Non-Ada tools and techniques much stronger than Ada counterparts |
| Personnel | • DOD has dominant position in Ada/applications skill base, which requires some investment to sustain<br>• Expensive for adversaries to create comparable Ada/applications skill base<br>• Non-Ada/applications skill base achievable, but with initial cost, lower competitive advantage | • Non-Ada/applications skill base much stronger than Ada counterpart |
| High assurance and real-time performance | • Ada superior; attributes are success-critical | • Ada superiority diluted by COTS; attributes are less success-critical |
| Ease of change | • Ada somewhat superior, more so for high-assurance changes | • Ada somewhat superior for custom software but at a disadvantage for COTS-based applications (see "Size" above) |
| Socio-technical infrastructure | • Existing DOD Ada-based infrastructure stronger than alternatives; requires some investment to sustain<br>• Expensive for adversaries to match and sustain comparable infrastructure | • Existing C/C++ -based infrastructure much stronger than Ada-based infrastructure |

In commercially dominated applications, software size can be significantly reduced by the use of commercial off-the-shelf (COTS) software and fourth-generation programming languages (4GLs). Given that most of these applications are developed in C and C++, and given current and likely future COTS volatility, it will generally be harder for DOD to use Ada, as compared to C and C++, to develop and sustain non-COTS portions of the software.

*Process*

As discussed in "Influence of Software Environments, Tools, and Languages on the Software Engineering Process" in Chapter 2, Ada enables the definition and verification of architectural interfaces early in the development process. This capability reduces a major source of rework, giving Ada another appreciable, but not overwhelming, advantage in the largely custom warfighting sector.

The more COTS-intensive, commercially dominated applications would not benefit much from Ada-based early interface checking (absent a significant investment in sustaining Ada interface specifications for each evolving COTS product). Further, COTS-based rapid development processes based on early availability of C and C++ interfaces would leave Ada at some disadvantage.

*Environment*

In warfighting applications, investment in very high assurance, real-time Ada software tools has been beneficial to DOD. Without continued investment, however, Ada tools will not be robust in either warfighting or commercial applications. The predominance of C and C++ tools in such areas as mainstream databases and graphical user interfaces places Ada at a significant disadvantage.

For all applications, the large commercial marketplace will continue to stimulate development of a large selection of general software tools (e.g., smart editors, "debuggers," static and dynamic analyzers, and tools for configuration management, quality assurance, and testing), more frequently supporting C and C++ than Ada. Ada parity for general tools would require additional DOD investment.

*Personnel*

For development of software for warfighting, DOD has in place an established group of experts who understand the existing base of Ada warfighting software and can expeditiously extend it. Some DOD investment in Ada-based education and training is necessary to sustain the current skill base and to add new personnel, but this investment is minor compared to the investment an adversary would need to make to compete across the board with DOD in warfighting software (either by building an Ada skill base or developing a comparable warfighting applications software base in another language). Competing in warfighting niches (e.g., information warfare and security) would still be feasible for adversaries, independent of the choice of programming language. Although DOD could retrain its warfighting software programmers to use C and C++, thus avoiding the need for education in more specialized programming languages, such an approach would require a considerable time delay as well as investment for retraining the existing Ada applications skill base, and it would eliminate the competitive advantage currently enabled by the Ada applications skill base.

Because commercially dominated applications software is written mostly in C, C++, and other non-Ada languages such as Cobol, Java, and 4GLs, DOD use of Ada in this sector would forego the advantages of relying on the base of commercial skills.

*High Assurance and Real-Time Performance*

As described in Appendix B and summarized in the Chapter 2 section titled "Technical Evaluation of Ada 95 and Other Third-Generation Programming Languages," Ada provides a significant technical advantage for achieving high assurance and real-time performance, two attributes that are critical to DOD's competitive advantage for many warfighting applications.

In commercially dominated applications, levels of assurance that can be provided by COTS represent the standard. Complementing COTS-based capabilities with those provided by Ada would not improve much on COTS levels of assurance and real-time performance, which in general are sufficient for all but a small fraction of commercially dominated applications.

### *Ease of Change*

As described in Appendix B and summarized in the Chapter 2 section "Technical Evaluation of Ada 95 and Other Third-Generation Programming Languages," Ada's encapsulation, type-checking, "generics," and other capabilities facilitate changes in software and lead to lower maintenance costs for custom software.  With Ada, high levels of assurance can be sustained for modified software, a distinct advantage for warfighting applications.

As discussed for the criterion of software size, requiring the use of Ada for COTS-dominated commercial applications would be a disadvantage because of the need to adapt to frequent COTS changes.

### *Socio-Technical Infrastructure*

DOD has developed a strong base of aerospace contractors and subcontractors who have in common a familiarity with both Ada and warfighting applications, enabling them to quickly combine in various ways to address new needs for and opportunities in warfighting software.  No DOD adversary has a comparable range of capabilities or is likely to make the major investment required to generate and sustain such capabilities.  Thus, although a moderate DOD investment is required to sustain Ada-based warfighting software, the competitive advantage gained in warfighting applications makes it more than worthwhile.

On the other hand, for commercially dominated applications, the pervasiveness of C and C++ in the socio-technical infrastructure places Ada at a major disadvantage.

### Conclusions

In weighing the factors discussed above and listed in Table 3.1, the most compelling points for Ada's use for warfighting applications are the strong competitive advantages accruing to DOD in the areas of personnel, high assurance and real-time performance, and socio-technical infrastructure. Evaluation of Ada with respect to software size, process, development environment, and ease of change does not produce a sufficient rationale for requiring Ada's use. With respect to these criteria, Ada offers some moderate advantages, but, in some cases (e.g., for general software tools), DOD investment is required just to achieve parity between Ada and other available programming languages.

Based on the points above, there is sufficient justification to continue the requirement to use Ada in warfighting applications—if DOD also commits to the associated necessary investment in sustaining the infrastructure for Ada.  The committee believes that an appropriate level of investment is roughly $15 million per year (detailed in Chapter 5); the economic analysis presented at the end of this chapter indicates that this level of investment is justified.

Considering that programming languages are the materials out of which software is built, this investment strategy is a familiar one for DOD.  It is analogous to investing in and encouraging the use of high-performance physical materials that provide DOD with competitive advantages for its weapon systems (see Box 3.1).  However, this observation is not meant to imply that Ada technology should be put on the Militarily Critical Technologies List and be subject to embargo; rather, DOD can gain an advantage by sustaining its Ada capability for warfighting systems.

Consideration of the factors listed in Table 3.1 for commercially dominated applications leads to the conclusion that a DOD requirement for Ada-based software in commercially dominated   applications

**BOX 3.1**
**Ada 95 as a DOD Materials Investment**

Suppose that DOD has been developing a new generation of materials for high-performance weapon systems. These materials have four main characteristics:

1. They are based on proven cost-effective development and performance with a previous generation of materials.
2. Their use requires special manufacturing and maintenance expertise. DOD and its supplier community have the largest source of such expertise, including tooling and components.
3. They provide the potential for DOD to create and maintain systems that will significantly outperform adversaries' weapon systems.
4. They require a continuing investment in materials research and development and in manufacturing technology of less than 0.1 percent of the cost of the weapon systems in which they will be deployed.

When such materials are physical materials that contribute to low-observable, high-strength, or high-temperature performance, DOD routinely makes such investments. The same rationale can be applied to investment in weapon systems software, in which programming languages operate as the counterpart of physical materials. Each programming language contributes to varying extents to a warfighting application's malleability, ability to withstand stress, and suitability for various operational environments.

For embedded weapon systems, the rationale for investment outlined above applies well to Ada 95, which has characteristics analogous to those described for new materials above.

1. Several studies (see Chapter 2) have indicated that Ada 83 (the previous programming "material") has enabled more cost-effective development and maintenance of embedded weapon systems software than have other languages.
2. The DOD software community has by far the largest source of people capable of programming in Ada 83, particularly for embedded weapon systems, and DOD's strongest embedded weapon systems software tools and components are based on Ada 83.
3. Ada 95 incorporates a decade of experience in ways that improve on Ada 83's capability to address real-time, distributed, and object-oriented systems.
4. Ada 83's capabilities and Ada 95's potential are products of a continuing DOD investment of more than $10 million annually for the past decade. A comparable investment is necessary to provide DOD with the continuing improvements in Ada 95 tools, components, infrastructure, and education required to keep DOD well ahead of any other nation's ability to produce or modify embedded weapon systems software.

If DOD chooses to implement its future embedded weapon systems software in other programming languages such as C or C++, it can still produce good systems. However, such systems would provide lower levels of assurance than those produced with Ada 95. Also, choosing another programming language would require a large DOD investment in moving its weapon system software components, manufacturing expertise, and maintenance expertise from Ada to C or C++. A further result of such a transition would be to make DOD's weapon system software components and expertise easier for adversaries to assimilate.

*continued on next page*

**Box 3.1—*continued***

If an adversary chose to build its embedded weapon systems software using C or C++ and commercial off-the-shelf (COTS) components, it would have some advantages in tooling and a large labor pool to draw from.  However, it would have several formidable problems in competing with DOD's weapon systems software.  Chief among these problems are the amount of new C and C++ software (a large fraction of 50 million lines) that would need to be developed to replace the current inventory of Ada code; the need to retrain the labor force in weapon systems applications; the mismatch between COTS software (e.g., "Windows") and real-time, high-assurance requirements; and the need to match DOD's expertise in such areas as reuse (via Ada's package specifications and generics), sensor-based control algorithms and data structures in Ada, and Ada's real-time scheduling capability.

is not justified.  For custom 3GL software in this sector, Ada is a strong candidate and should be considered in programming language decisions.  However, it should not be necessary to provide extra justification for the use of languages other than Ada.

## FINDINGS AND RECOMMENDATIONS

Based on its assessment of today's environment for software development and its evaluation of DOD's current programming language policy (Chapter 1), its examination of trends in software engineering and comparison of various programming languages (Chapter 2), and the results of its business-case analysis to evaluate Ada in two software application domains (the first two sections of this chapter), the committee developed the following set of findings and recommendations for DOD.  The recommendations address the use of Ada in warfighting software, the application in which the committee finds Ada to have demonstrated benefit; the proper scope and implementation of DOD software policy; investment in Ada; and collection of data as a basis for assessing the effectiveness of software and software policy.

### Ada Competitive Advantage

**Finding**.  Ada gives DOD a competitive advantage in warfighting software applications, including weapon control, electronic warfare, performance-critical surveillance, and battle management.

**Recommendation**.  Continue vigorous promotion of Ada in warfighting application areas.

**Rationale**.  Available project data and analyses of programming language features indicate that, compared with other programming languages, Ada provides DOD with higher-quality warfighting software at a lower life-cycle cost.  DOD can increase its advantage by strengthening its Ada-based production factors (involving software tools, technology, and personnel) for warfighting software (see Chapters 2 and 3).

### Applicability of Policy to DOD Domains

**Finding**.  DOD's current requirement for use of Ada is overly broad in its application to all DOD-maintained software.

**Recommendation**. Focus the Ada requirement on warfighting applications, particularly in critical, real-time applications, in which Ada has demonstrated success. For commercially dominated applications, such as office and management support, routine operations support, asset monitoring, logistics, and medicine, the option of using Ada should be analyzed but should not be assumed to be preferable.

**Rationale**. For warfighting software, supporting Ada-based production factors (involving software tools, technology, and personnel) gives DOD a competitive advantage. In this domain, eliminating the use of Ada would both compromise this advantage and diminish the capabilities for maintaining DOD's existing 50 million lines of Ada. In commercially dominated areas, pushing applications toward Ada would create a disadvantage for DOD (see Chapters 2 and 3).

## Scope of Policy

**Finding**. DOD's current requirement for use of Ada overemphasizes programming language considerations.

**Recommendation**. Broaden the current policy to integrate the choice of programming language with other key software engineering concerns, such as software requirements, architecture, process, and quality factors.

**Rationale**. The current policy isolates the Ada requirement and the waiver process from other software engineering decisions, causing programs to make premature or non-optimal decisions (see Chapter 1). DOD has already taken steps to broaden the policy focus in its draft revision of its programming language policy, DOD Directive 3405.1; this report recommends modifications to that draft policy (Appendix A).

## Policy Implementation

**Finding**. DOD's current Ada requirement and the related waiver process have been weakly implemented. Many programs have simply ignored the waiver process. Other programs make programming language decisions at the system level, but often a mix of Ada and non-Ada subsystems is more appropriate (see Chapter 1).

**Recommendation**. Integrate the Ada decision process with an overall Software Engineering Plan Review (SEPR) process. Passing such a review should be a requirement for entering the system acquisition Milestone I and II reviews covered by DOD Instruction 5000.2. It should also be required for systems not covered in 5000.2, and recommended for DOD-directed software development and maintenance of all kinds.

**Rationale**. The SEPR concept is based on the highly successful commercial architecture review board practice. The SEPR process involves peer reviewing not only the software and system development plans, but also the software and system architecture (building plan) and its ability to satisfy mission requirements, operational concepts, conformance with architectural frameworks, and budget and schedule constraints; the process also involves reviewing other key decisions such as choice of programming language (see Chapter 4).

## Investment in Ada

**Finding**. For Ada to remain the strongest programming language for warfighting software, DOD must provide technology and infrastructure support.

**Recommendation**. Invest in a significant level of support for Ada, or drop the Ada requirement. The strategy developed by the committee recommends an investment level of approximately $15 million per year.

**Rationale**. With investment, DOD can create a significant Ada-based complex of production factors (involving software tools, technology, and personnel) for warfighting application domains. Without such support, Ada will become a second-tier, niche language such as Jovial or CMS-2 (see Chapter 5).

### Software Metrics Data

**Finding**. DOD's incomplete and incommensurable base of software metrics data weakens its ability to make effective software policy, management, and technical decisions.

**Recommendation**. Establish a sustained commitment to collect and analyze consistent software metrics data.

**Rationale**. The five sets of findings and recommendations above are based on a mix of incomplete and incommensurable data, anecdotal evidence, and expert judgment. For this study, the patterns of consistency in these sources of evidence provide reasonable support for the results—but not as much as could be provided by quantitative analysis based on solid data. A few organizations within DOD have benefited significantly from efforts to provide a sound basis for software metrics; a DOD-wide data collection effort would magnify the net benefits (see Chapter 2).

### ASSESSMENT OF POLICY ALTERNATIVES

Table 3.2 summarizes current DOD policy, the committee's recommended alternatives to the policy currently in force, and other programming language policy alternatives suggested to or considered by the committee. The subsections that follow in the text summarize the committee's rationale for preferring these recommended actions over the alternatives.

### Conditions for Requiring Ada

There are five conditions to be examined in determining whether a software development is subject to the Ada requirement. The following subsections describe the approach to defining these conditions. The committee recommends that *all* of these conditions be met in order for Ada to be required.

Table 3.2 Policy Alternatives Recommended by the Committee and Alternatives Considered

| Policy Item | Current Policy | Committee Recommendation | Alternatives Considered |
|---|---|---|---|
| **CONDITIONS FOR REQUIRING ADA** | | | |
| Application domain | All DOD software systems; all sectors | Application subsystem is in the warfighting sector | • None<br>• Other subsets |
| Maintenance | DOD-directed maintenance | Same as current policy | • All DOD software<br>• Only DOD-performed maintenance |
| Level of applicability | Entire system | Subsystem is critical or is larger than 10 KLOC | • Other size, criticality criteria |
| Existing solutions | Consider COTS, NDIs 4GLs | No better COTS, NDI, or 4GL solution exists | • Require COTS, NDIs, 4GLs |
| Other languages | No life-cycle cost-effectiveness justification for use of another language | Same as current policy | • Require another language (C, C++, Java)<br>• Require object-oriented solutions |
| **ADA REQUIREMENT** | | | |
| System coverage | Entire system; 100% Ada code | 95% or more of subsystem in Ada | • Other percentages |
| **LANGUAGE CHOICE PROCESS** | | | |
| Exceptions | Waiver-based; independent of other software engineering reviews | Part of Software Engineering Plan Review process | • Tied to different software engineering reviews |
| Approval | SAEs or ASD (C$^3$I) | Approval delegated through SAEs to appropriate level; periodically reviewed by ASD (C$^3$I) | • ASD (C$^3$I) only<br>• Delegated to project manager's superior; periodically reviewed by SAEs, ASD (C$^3$I) |
| **INVESTMENT IN ADA INFRASTRACTURE** | | | |
| Level | $10 M in FY94; $0 M in FY98 | $15 M/year | • $2 M/year (barely sustaining)<br>• $30 M/year (major initiative) |

NOTE: ASD (C$^3$I), Assistant Secretary of Defense (Command, Control, Communications, and Intelligence); COTS, commercial off-the-shelf; 4GL, fourth-generation programming language; KLOC, 1,000 lines of code; NDIs, non-developmental items; SAE, Service Acquisition Executive.

*Application Subsystem Is in the Warfighting Sector*

Current DOD policy requires that Ada be used for all applications (subject to a waiver process and the criteria in Table 3.2). The committee recommends requiring Ada only for subsystems in DOD-dominated warfighting applications, as defined above in this chapter. For other DOD applications, Ada should be considered as a candidate but should not be required. The focus on subsystems makes it possible to separate the weapon control aspects of a system, which should be written in Ada, from other aspects of the system, which may be written in Ada, but need not be. An aircraft carrier is a good example of such a system.

As discussed above, in non-warfighting applications, the rapid pace of non-Ada commercial software solutions shifts the business case away from Ada. For warfighting applications, a focused Ada strategy can generate a differential that enables DOD to develop and adapt software more readily than its adversaries.

The primary alternative to be considered is to not impose any requirement. Advocates of this approach (AIA, 1996) argue that developers of warfighting software can choose a programming language by analyzing trade-offs, in the same way that they make decisions about other non-developmental items (NDIs), such as commercial hardware and COTS software. The committee recognizes that many DOD and contractor organizations can make good technical choices when given the freedom to do so. However, the uneven distribution of software expertise across the DOD acquisition system could lead to some poor results, therefore leaving DOD worse off under an open policy than it would be under a requirement to use Ada in specific critical applications.[4]

The primary problem with leaving programming language selection to developers is that DOD acquisitions still are done largely one system at a time, rather than according to a product-line approach. Currently the focus in source selection is the cost of developing individual systems, and less attention is paid to important factors, such as assurance and adaptability, that are difficult to quantify. Such an approach can frequently lead to situations in which bidders can propose cheap but fragile solutions (e.g., using immature commercial-grade non-Ada process control software). DOD's program management staff often does not have sufficient software expertise (see Chapter 1) to be able to disqualify such inadequate solutions. The net result for DOD warfighting software then becomes a mix of strong and weak Ada and non-Ada solutions, and a socio-technical infrastructure characterized by increasing problems with proliferation of programming languages. Requiring Ada in warfighting applications is not a panacea, but it should lead to better-performing systems for DOD.

Other possible policy alternatives could require that Ada be used for other subsets of DOD software, such as all software maintained by DOD personnel. In this particular case, Ada's use would be required for a great deal of financial and other commercially dominated software, which the committee's business-case analysis indicates would not be a cost-effective approach for DOD. The committee was unable to identify other subsets in which mandating use of Ada would match the cost-effectiveness of requiring its use in warfighting applications.

*Maintenance Is Directed by DOD*

DOD's current requirement for use of Ada excludes software that is embedded in a warfighting system component (e.g., an altimeter) purchased as a supplier-maintained commercial item. In such cases, DOD can frequently capitalize on dual-use (commercial and military) products as it does on COTS software. The committee recommends maintaining this aspect of the policy. The alternative of requiring Ada for all software in DOD warfighting systems is less cost-effective because it cuts off the opportunity to use many of these dual-use components. Another alternative, requiring use of Ada only for DOD-performed rather than DOD-directed maintenance (i.e., maintenance performed by DOD as well as by its contractors) would invite proliferation of programming languages in DOD's contractor-maintained warfighting software inventory, with the same consequent relative weakening of DOD's position discussed in the previous subsection.

*Subsystem Is Critical or Larger Than 10,000 Lines of Code*

The committee concluded that DOD's current policy of requiring waivers for use of non-Ada software for small, non-critical subsystems can cause bureaucratic delays that add little value. DOD should consider alternative size and criticality criteria. In suggesting alternatives, the committee focused on the objective of keeping the policy simple. The metric of 10,000 lines of code is based solely on engineering judgment: the committee's consensus was that 3,000 lines was too low and 30,000 lines was too high. "Critical" should be interpreted as key to mission success.

*No Better COTS, NDI, or 4GL Solution Exists*

As discussed in Chapter 1, the current DOD policy establishing Ada's precedence over COTS, NDI, and 4GL solutions has led to a number of bureaucratic delays or incorrect choices of an Ada solution over a more cost-effective COTS solution.

In addition to its recommended alternative, another alternative considered by the committee was to require the use of COTS, NDI, and 4GL solutions. As discussed in Chapter 2, these solutions can have many drawbacks for DOD and other systems, and requiring their use would push many DOD systems into adopting less cost-effective solutions.

*No Life-cycle Cost-effectiveness Consideration Justifies Use of Another Language*

This criterion for required use of Ada is identical to DOD's current policy. As recommended below, embedding of the Ada waiver process within the SEPR process (Chapter 4) should make justifications for use of non-Ada software both sounder and less onerous.

Alternatives suggested to the committee were to require the use of other programming languages (e.g., C, C++, or Java) or to require the use of object-oriented approaches. The comparisons of language features in Chapter 2 and the business-case analysis presented above in this chapter do not support a recommendation for requiring other languages. With respect to object-oriented techniques, the committee's assessment was that these are not well enough defined or sufficiently mature to support a requirement for use.

## Ada Requirement

If all of the above criteria hold for a given software development, the committee's recommendation is that the requirement to use Ada should hold. The following subsection analyzes the appropriate coverage of the requirement.

*95 Percent or More of the Subsystem's Warfighting Software Is to Be Written in Ada*

The current DOD policy calls for the entire system to be developed in Ada. The committee recommends only that at least 95 percent of the applicable subsystem use Ada. Again, the intent of this recommendation is to avoid bureaucratic delays in cases involving small amounts of non-Ada "glue code" or performance enhancements, and 95 percent is again based on engineering judgment. The committee believes that 90 percent is too low and 98 percent too high. As with the criterion for subsystem size above, the scope of this requirement could be subject to some manipulation by program desire to evade the requirements.  However, the committee concluded that coupling the Ada waiver process with the SEPR process detailed in Chapter 4 would eliminate most of the gamesmanship in a cost-effective manner.

## Language Choice Process

The following subsections recommend alternative approaches to implementation and approval of the language choice process.

### *Replace the Waiver Approval Process with Other DOD Software Reviews*

The current DOD Ada waiver approval process is disconnected from other reviews of a project, causing extra work and out-of-context decisions (see Chapter 1).  As discussed in Chapter 4, embedding the waiver process within the SEPR process should make it sounder and less onerous.

Requests for waivers could also be considered at other review and approval milestones.  The committee's recommendation associates the SEPR reviews with the Major Automated Information Systems Review Council (MAISRC) reviews at Milestones I and II, but recommends that the SEPR reviews for major weapon systems be completed as prerequisites to the counterpart Defense Acquisition Board (DAB) Milestone I and II reviews (see Figure 3.1).  The SEPR reviews could be embedded within the DAB reviews, but experience to date indicates that software issues have had little visibility in many previous DAB reviews.  The SEPR reviews and programming language decisions could also be handled in conjunction with source selection, but source selection may occur either too early or too late to allow effective consideration of waiver requests, depending on the relative timing of source selection and DAB/MAISRC milestone reviews.  Basing a SEPR review on a 60-day proposal effort could also make it too superficial and vulnerable to manipulation.  On the other hand, requiring an SEPR review for source selection would likely stimulate better proposals.

### *Reconsider the Level at Which Waivers Can Be Approved*

The current DOD waiver policy requires Service Acquisition Executive (SAE) approval for service-level waivers and Assistant Secretary of Defense ($C^3I$) approval for waivers in joint programs.  As discussed in Chapter 1, this requirement for approval at extremely high levels has induced considerable avoidance of the waiver process (sometimes leading to inappropriate use of Ada, sometimes simply to disregarding of the requirement).  The committee recommends that SAE-level approval via the SEPR process be required only for major system acquisition, and that authority for granting waivers be delegated, as part of the SEPR process,  to the equivalent of a product-line manager (a Program Executive Officer for families of warfighting systems; a base commander for base-specific systems).  This approach would make the reviews and waiver process more mission-relevant.  Periodic reviews by SAEs and the Assistant Secretary of Defense ($C^3I$) to assess the effectiveness of the SEPR and Ada waiver process would ensure their long-range effectiveness without involving top executives regularly in low-level system decisions.

Waiver approval at levels even higher than the current level (e.g., that of the Assistant Secretary of Defense ($C^3I$) for every project) could also be considered but would exacerbate the current problems of critical-path delays and avoidance of the waiver process.  Delegation of authority for approval to even lower levels (e.g., the project manager's direct superior) could also be considered, but in many cases managers at such levels do not have sufficient visibility or responsibility across product-line objectives to make appropriate decisions affecting long-range product-line effectiveness.

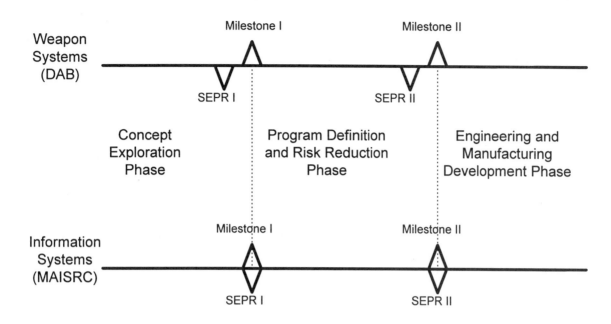

FIGURE 3.1  Integration Software Plan Review (SEPR) process with Defense Acquisition Board (DAB) and Major Automated Information Systems Review Council (MAISRC) milestones for weapon and information systems.

## Investment in Ada Infrastructure

The current DOD strategy is to reduce investment in Ada from a $10 million annual budget in Fiscal Year 1994 to essentially zero in Fiscal Year 1998.  Other programming languages that DOD uses but provides little or no support for include Jovial and CMS-2; at this level of investment, bindings, run-time systems, language features, and tools are kept current only by projects using the language.  This level is inadequate to keep a programming language up to date and serves as a major disincentive for projects to continue to use the language.  Thus, DOD's planned zeroing out of investment in Ada is incompatible with any DOD requirement to use Ada.

Chapter 5 identifies a set of investments that are necessary to provide DOD with a competitive advantage in warfighting software, and indicates that its magnitude is roughly $15 million per year.  The economic analysis in the following section indicates that the resulting cost savings justify this level of investment.

An alternative policy would be to attempt to sustain Ada at a considerably lower level, on the order of $2 million annually.  This level would be enough to enable DOD to keep its current 50 million lines of Ada from becoming hopelessly obsolete, but not enough to keep new programs from deciding that Ada would require too much extra investment to justify its use.  Another alternative would be a major initiative level of investment, at roughly $30 million per year.  This was the level recommended by the 1991 DOD software technology strategy for a 3-year period (DOD, 1991).  It included such additions as a demonstration/validation phase for Ada 95, a major suite of tools for re-engineering legacy software into Ada, and investment in beyond-Ada 95 programming language research.  Since 1991, the primary window of opportunity for an Ada 95 demonstration/validation phase has passed; programming language research is now being pursued in general by the National Science Foundation and in particular by the Defense Advanced Research Projects Agency's Dynamic Object-Oriented Language initiative.  Given this situation, a strategy of investment at the margin in such areas as re-engineering tools and real-time, high-assurance tools at the $15 million annual level appears more cost-effective than a major initiative, and can be justified given the large installed base of Ada warfighting software.

## ECONOMIC ANALYSIS OF INVESTMENT IN ADA INFRASTRUCTURE

Given the fragility of the underlying data, a complex economic analysis attempting to establish the optimal investment in Ada infrastructure is not justified.  Instead, a simple analysis is presented to estimate how the recommended $15 million level of investment will affect Ada tool capability, tool experience, and language capability (see Box 3.2). The simplifying assumptions, all made on the conservative side, are as follows:

1. *Only the annual maintenance of the 50 million lines of Ada code in existing weapon systems software is considered.*  This assumption is conservative because it does not consider developing and sustaining current and future Ada systems.

2. *Only 8 percent of the software will be changed per year.*  This assumption is conservative with respect to the average rate of change of 11 percent per year for embedded control software and 20 percent per year for human-machine interactive software in the COCOMO database (Boehm, 1981).

3. *The cost per changed line of code is only $40.*  This assumption is conservative with respect to general estimates that changing weapon systems software costs at least $75 per line of code.

4. *Only the effects of investment on Ada tool capability and Ada language and tool experience are considered.*  This assumption is conservative with respect to other cost effects due to process, reuse, and size improvements (separate vs. common development of infrastructure software).

5. *The presence versus the absence of an investment program for Ada will affect improvement in tool capability and language and tool experience by one cost model rating level (see Box 3.2 for details).*  This assumption is conservative with respect to the relative effects on other languages that are not supported by such investment, such as Jovial and CMS-2.

6. *The effect of increasing tool capability and language and tool experience one rating level is an 8 percent improvement.*  This assumption is conservative with respect to most estimates generated by software cost models, which indicate that the improvement per rating level for these variables averages 10 to 12 percent (see Box 3.2).

Given these assumptions, an estimate of the annual cost of sustaining the 50 million lines of code (50 MLOC) of existing DOD weapon systems software in Ada is

$$(50 \text{ MLOC}) (0.08/\text{year}) (\$40/\text{LOC}) = \$160 \text{ million/year}.$$

With a one-level improvement in tool capability and language and tool experience, the estimated annual cost of sustaining existing DOD weapon systems software in Ada is

$$(\$160 \text{ million/year}) (0.92) (0.92) = \$135 \text{ million/year}.$$

Thus, a conservative estimate of the annual maintenance cost savings resulting from a one-level improvement in capability is $25 million per year, considerably more than the annual investment in Ada of $15 million per year that would be required to achieve the improvement in rating level.

## BOX 3.2
## Software Cost Models and the Effects of Investment

As discussed in Chapter 2, most current models for estimating software costs have input parameters or cost drivers reflecting the product's size and the project's personnel, development environment, and process characteristics. The personnel and environment cost drivers generally take the form of multipliers affecting the estimated cost. For example, the COCOMO 2.0 TOOL model rates cost drivers for tools and corresponding cost multipliers as shown in the following table.

### Relative Estimated Cost for Various Levels of Tool Support

| Support Level | TOOL Rating Level | Cost Multiplier |
|---|---|---|
| Very Low | Edit, code "debugging" tools | 1.20 |
| Low | Simple front-end, back-end CASE; little integration | 1.10 |
| Nominal | Basic life-cycle tools; moderately integrated | 1.00 |
| High | Strong, mature life-cycle tools; moderately integrated | 0.88 |
| Very High | Strong, mature, proactive life-cycle tools; well integrated with processes reuse | 0.75 |

The stronger the project's tool support, the lower its estimated cost. The estimated cost of a project with low tool support (simple front-end and back-end computer-automated software engineering (CASE) tools with little integration) is 10 percent higher than the estimated cost of a project with nominal tool support (basic life-cycle tools with moderate integration). The values in the table are based on a mix of several organizations' experience and calibration to project data. COCOMO 2.0 is currently calibrated with data from 65 projects, for which the calibrated cost estimates are within 25 percent of the actual cost 60 percent of the time.

The corresponding rating scale and cost multipliers by the COCOMO 2.0 LTEX (language and tool experience) model are shown immediately below.

### Relative Estimated Cost for Various Levels of Support for Programming Language and Tool Experience

| Support Level | LTEX Rating[a] | Cost Multiplier |
|---|---|---|
| Very Low | 2 months or less | 1.24 |
| Low | 6 months | 1.11 |
| Nominal | 1 year | 1.00 |
| High | 3 years | 0.90 |
| Very High | 6 years | 0.82 |

[a] LTEX: Language and tool experience

*continued on next page*

**Box 3.2—*continued***

The use of calendar time as a basis for evaluating experience has both the advantage and disadvantage of being simple.  The estimated cost of a project using personnel with low (median of 6 months) language and tool experience is 11 percent higher than the cost of using personnel with a nominal (median of 1 year) level of experience.

Other models estimate similar costs.  For example, the Jensen-SEER model (Jensen and Lucas, 1983) and the Softcost-R model[1] have cost drivers comparable to the COCOMO TOOL and LTEX cost drivers, with cost differentials in the range of 10 to 12 percent per rating level.  Checkpoint has a "CASE tools" cost driver, with cost differentials varying from 10 to 15 percent per rating level, depending on interactions with other cost drivers covering staffing, methods, and language level (Jones, 1996d).

A recent sample of 25 Ada projects from the Air Force Cost Analysis Agency indicates an average TOOL rating for the level of tool support of "nominal" and an average LTEX rating for the level of language and tool experience support of "high."  As discussed in Chapter 5, the committee's recommended strategy for DOD's investment in Ada tools is to invest at the margin with tool suppliers to develop tools with clear benefits that otherwise would not be built.  With this investment, the level of Ada tool support likely would increase by one TOOL rating level over the next 5 years.  Without this investment, the level of Ada tool support would likely decrease with respect to the pace of surrounding technology.  Thus, it is reasonable to assume that the net effect of the Ada investment would be an improvement of one level on the TOOL rating scale, and a corresponding 8 percent cost reduction.

Another effect of the committee's recommended Ada investment strategy, discussed in Chapter 5, would be to stimulate education and training in Ada, again via collaborative investments with suppliers at the margin.  Without this investment, it will be much harder 5 years from now to find sources of qualified Ada personnel, and the likely net effect will be a one-level decrease in the average LTEX rating for Ada projects.  With the Ada investment, the likely effect will be an increase in the average rating.  Thus, a conservative assessment of the net effect of the committee's recommended Ada investment would be an increase of one level on the LTEX rating scale, and a corresponding 8 percent cost reduction.

---

[1] D.J. Reifer, personal communication to B. Boehm, August 1996.

**NOTES**

1. See also Perry (1996b).

2. This process is illustrated by the experience of Xerox as described in Weidernan (1991).  Xerox's Digital Systems Department evaluated Ada, C++, and two other languages for use in large, real-time, embedded systems software, and concluded that Ada was superior in terms of language features, implementation, and cost.  In practice, however, the bulk of Xerox's embedded systems software continues to be developed in C and C++.

3. For example, Gerald Pasternack of Citicorp testified to the committee about low awareness of and a low supply of Ada programmers in the New York financial market.

4. The committee heard from several DOD representatives that measures to reform acquisition emphasize performance requirements over military specifications.  This trend can be interpreted as supporting the option of having no requirement to use Ada.

# 4

# Implementation of Recommended DOD Software Policy

The committee's recommendations for DOD's software policy address two broad objectives. The first part of this chapter describes appropriate principles for selection of a programming language, and Appendix A contains the committee's proposed modifications to a revised version of DOD Directive 3405.1 (DOD, 1987a), which was in the process of being redrafted during the course of this study. The second component of the committee's recommendations concerns the Software Engineering Plan Review, which is proposed as a method for implementing DOD's software policy and is described in the second part of this chapter.

## RECOMMENDED POLICY FOR CHOICE OF PROGRAMMING LANGUAGE[1]

The committee recommends that DOD approach programming language policy at three levels of precedence. The overall goal is to achieve the best combination of costs and benefits (each interpreted quite broadly, as explained below); a number of principles for acquisition of software follow from and are subordinate to this overriding goal. The second level of precedence interprets those principles as they apply to the choice of a programming language (at any level of programming). The third level specifies circumstances under which Ada is required for software development using a third-generation programming language (3GL).

This hierarchy expresses goals for software acquisition that are broader than the choice of programming language alone, clarifying the importance of many other decisions (such as decisions about whether to make, buy, or, build components; design of the development process; and necessary skills) required to achieve DOD's goals.

The focus is on operational software. It does not apply to software developed, acquired, or used by DOD research and development activities, funded by 6.1, 6.2, and 6.3a appropriations. However, research and development software efforts likely to lead to new DOD operational capabilities should include plans for the transition of such software to meet operational software policy requirements (these plans are described under "Approval Authority and Milestones" in the next section).

## Goals of Software Development

High quality, low cost, and timely delivery are the primary goals for software development. Here, "quality" and "cost" are interpreted broadly. Quality includes, but is not necessarily limited to, functionality, fitness for a purpose, assurance (including reliability, survivability, availability, safety, and information security), efficiency, ease of use, interoperability, future adaptability (including extensibility, maintainability, portability, scalability, and compliance with standards), and development of DOD's software expertise. Cost includes, but is not limited to, full life-cycle monetary costs (i.e., both short- and long-term costs) and the extent of use of other scarce resources such as expert personnel. Cost also includes assessment of program risk and monetary and non-monetary consequences of system failure. Timely delivery, or schedule, is listed as a third goal because it is difficult to classify as either a quality or cost factor. These overriding goals are reflected in the following statements, which the committee believes should serve as guidance for DOD software development.

1. *Projects will specify and prioritize quality, cost, and schedule goals, and will analyze trade-offs and the business case for particular decisions.* Failure to articulate and prioritize project requirements appropriately, and to analyze them in the context of their impacts on cost and schedule, commonly leads to project failure or inappropriate acquisition decisions. It is not reasonable for DOD to specify a single prioritization of goals, because the importance and relevance of different factors vary widely. However, projects should conduct an analysis and defend it in the review process. Requirements should not be overstated, an approach that often has the effect of ruling out simpler, more cost-effective solutions.

2. *Projects will not develop new software unless quality, cost, and schedule goals cannot be met with non-developmental items (NDIs).* Developing and maintaining new software within projects tend to be more expensive than reusing suitable existing software.

3. *Commercial items are preferred over other non-developmental items if they meet quality, cost, and schedule constraints.* True commercial items will spread the costs of maintenance and improvement over a larger base, leading to cost savings. Issues such as possible "lock-in" to a single source should be considered as constraints to achieving desirable qualities such as adaptability and portability.

4. *Software development will emphasize good software engineering practice,* including the application of management techniques, methodologies, support tools, metrics, and appropriate programming languages. Good practices provide better quality at lower cost, regardless of which programming language is used. Good practices also tend to improve timeliness and reduce risk.

5. *Software developers should be chosen based on their experience,* a criterion that includes, but is not limited to, successful past performance; experience in the software domain or product line; use of appropriate management techniques, methodologies, support tools, and metrics; and mature software engineering capability and expertise.

6. *Projects will, when possible, exploit and/or contribute to open system architectures and common product lines, frameworks, and libraries.* Investment in commonality, where feasible, increases portability and opportunities for reuse, and reduces cost.

7. *Projects will avoid developing project-specific tools and technologies unless the cost, schedule, and/or quality advantage can be defended.* Such development is expensive and is seldom justified.

## Guidelines for Choice of Programming Language

1. *Projects will use the highest-level language that meets quality, cost, and schedule constraints for each software component.* Other things being equal, higher-level languages increase productivity and reduce cost. Specifically, 3GLs (high-order languages) are generally preferable to machine or assembly language; further, fourth-generation programming languages (4GLs), program generators, graphical user interface builders, and database query languages, such as Structured Query Language (SQL), are generally preferable to 3GLs. Modification of the lower-level language output from a higher-level language processor should be considered as programming at the lower-level; that is, components written in a language should be maintained in that language, and the output of a language processor should be changed only in exceptional cases.

2. *Standardized and non-proprietary languages are preferred.* Using standardized languages increases the portability of code and programmers, and diminishes the possibility of "lock-in" to a single source. This principle applies at all language levels. Thus standard SQL is preferable to a proprietary database query language. In some cases, unusual or "niche" languages are the best choice; however, these choices need to be defended.

3. *Projects should not develop new languages, and language processors for them, except for domain-specific languages that provide directives for application generators.* Such development is costly, in both the short and long term, and should require unusual justification.

4. *All relevant quality, cost, and schedule factors should be considered in the choice of programming language for each component.*

Applying these four principles, it is reasonable, for example, to use small "shell" scripts to "glue" together system components, rather than writing them in Ada or some other high-order language; however, large and complex shell scripts may violate the principles by being difficult to maintain. Likewise, large packages of spreadsheet macros, or other code written in (more or less) proprietary 4GLs, need to be considered carefully. The key is to ensure that decisions are made carefully, weighing all relevant economic and engineering cost, quality, and schedule factors. These requirements lead to the following recommended policy for the use of Ada.

## Recommended Policy for Requiring the Use of the Ada Programming Language

The committee believes that Ada should be presumed to be the best choice, and thus should be used for software development, for subsystems of DOD's operational software systems that meet **all** of the following criteria:

1. *The subsystem is in a warfighting software application area, as defined in Chapter 3.* While Ada may still be a good choice for other systems, DOD policy should require that Ada be used only in areas where it has clear advantages and is most likely to maximize DOD's competitive position relative to that of its adversaries.

2. *DOD will direct the maintenance of the software.* If a vendor is serving a broader customer community, then maintenance costs are spread over a larger base and are thus of less concern. If DOD directs the maintenance, whether or not the maintenance is performed by DOD personnel or a vendor, then DOD must cover the life-cycle cost, and Ada is assumed to be more cost-effective over an entire life-cycle.

3. *The software subsystem is large, more than 10,000 lines of code, or the subsystem is critical.* Small and non-critical subsystems, as a rule, incur lower development and maintenance costs, and thus

are not worth the cost of oversight. Such systems tend to be simpler and the choice of programming language is less critical. However, the choice is *more* important for critical components.

4. *There is no better COTS, NDI, or 4GL software solution.* If existing software or higher-level language solutions are suitable, new development solely to promote Ada should not be required.

5. *There is no life-cycle cost-effectiveness justification for using another programming language.*

6. *New software is being developed or an existing subsystem is being re-engineered; a re-engineering is a modification substantial enough that rewriting the subsystem would be cost-effective.* For systems meeting criteria 1 through 5, Ada is generally superior to other high-order languages, and conversion over time should be encouraged.

For systems that meet all of the above criteria, *Ada (preferably Ada 95) must be used for the preponderance (95 percent) of new or modified software subsystems or components; up to 5 percent may be written in other languages to facilitate component integration and other functions.*

Projects that meet all of the criteria except number 1 above must analyze Ada as an alternative. As explained in Chapter 2, Ada is generally preferred for custom software because, compared with other 3GLs, it encourages better software development practices, has better error checking and recovery capacity, has better support in certain domains, is standardized and has a validation facility, contributes to commonality, and leads to high quality at lower life-cycle cost.

## SOFTWARE ENGINEERING PLAN REVIEW PROCESS

The committee recommends that DOD broaden its current policy on programming language to include a range of software engineering factors that have a greater overall influence on software capability than does choice of a particular language alone. This section addresses how policy guidance regarding these factors, as described in Chapter 2, can be translated into operational decisions in systems development. The principal mechanism is the Software Engineering Plan Review (SEPR).

The committee explored a number of approaches for integrating selection of a programming language with related review and approval processes for software engineering decisions. One approach was to integrate the programming language selection process with a Capability Maturity Model assessment (Paulk et al., 1993), but this type of assessment focuses more on organizational process maturity than on specific technical decisions made by a particular project. Another approach was to add review of programming language and software engineering decisions to the Defense Acquisition Board (DAB) and Major Automated Information Systems Review Council (MAISRC) Milestones I and II review processes (as defined by DOD Directive 5000.2-R (DOD, 1996c)). Key software decisions are generally covered well in MAISRC reviews but often fall below the threshold of visibility in DAB reviews, which cover most DOD-dominated software application areas.

The committee determined that the DOD's best alternative to these two approaches was to require passage of a focused SEPR as a part of a major system's DAB or MAISRC Milestone I and II reviews. SEPRs used in commercial practice have proven to be highly effective for reviewing software and system requirements, plans, architectural decisions, and programming language decisions at life-cycle points similar to DAB and MAISRC Milestones I and II. For example, the SEPR concept has been used successfully in large technology-dependent commercial and government organizations, including AT&T and Lucent Technologies (architecture review board (AT&T, 1993)), Citibank (building permit system), NASA (architecture reviews), and others.[2]

The SEPR process is intended to provide a forum for the following activities:

1.  Involvement of stakeholders in key software engineering decisions,
2.  Contributions of peers and experts to key software engineering decisions,
3.  Stimulating commonality of process and architectural elements where appropriate, and
4.  Establishing accountability of a senior acquisition official for major software engineering decisions throughout the life cycles of related systems.

The SEPR process is intended to help program managers (and possibly contractors) achieve a best-practices level of decision making for the software engineering associated with major systems, as well as to assure consideration of organizational and life-cycle factors.  Implementation details are established not by senior officials, but rather by product-line stakeholders and expert peers, who have an incentive to minimize unnecessary bureaucracy and documentation.  The principal policy elements for systems subject to DAB and MAISRC reviews are the following:

1.  Authority for approving software engineering plans resides in the office of the Assistant Secretary of Defense ($C^3I$) and in the Service Acquisition Executives (SAEs) and their Software Executive Officials (SEOs).
2.  SEPRs are conducted by Software Engineering Plan Review Boards (SEPRBs) at key points in the engineering process and are conducted by peers and representatives of key stakeholders. These reviews are typically managed at the Program Executive Officer (PEO) level.
3.  Software engineering plans, focusing on major software engineering process, technology, and architecture decisions,  are submitted by program managers in preparation for the SEPR  process.
4.  The Office of the Assistant Secretary of Defense ($C^3I$) periodically reviews the effectiveness of the DOD services' and DOD components' implementation of the SEPR process.

The SEPR process has three elements:  (1) a policy framework established for major software engineering decisions and for SEPRs;  (2) involvement in the review by peers as well as the principal stakeholders in system design; and (3) software engineering common practices, SEPR evaluation criteria, and SEPR process policies developed at the service and command levels (these would be specific to each service, and possibly to PEOs who could, for example, require conformance to particular architectural frameworks for a class of systems (e.g., a particular level of a common operating environment).  These three elements are detailed below.

## Policy Framework

The purpose of the SEPR process is to embody institutional and long-term interests in requirements for formulation, development, and post-deployment that might otherwise be neglected or compromised in favor of short-term goals.  Such short-term expedients could arise as undesired results of incentives created in the acquisition process or for other reasons.

Early decisions concerning design, process, and other software engineering factors can have a significant influence on overall life-cycle cost and risk, and on the potential for product-line commonality and interoperability.  For example, the following questions arise:

• What is the necessary level of maintainability (e.g., ongoing improvements in performance and quality, and evolution of computational infrastructure), and how will it be achieved?

• What is the necessary level of interoperability (e.g., within product lines, with related DOD systems, and with related systems controlled by allies and in coalition forces), and how will it be achieved?

- What is the necessary level of trustworthiness (including reliability, fault tolerance, and survivability), and how will it be achieved?
- What are the likely future needs (e.g., new and changed requirements anticipated), and how will they be accommodated?
- What are the likely technology constraints, and what plans have been developed for inserting new technology?


## Stakeholder Role

The committee recommends that the SAEs be in charge of carrying out the SEPR process at the DOD service level. The SAEs would establish milestones for the SEPR process, appoint expert reviewers and stakeholder representatives, and establish criteria for evaluation. The SAEs and their associated SEOs would be responsible for implementing these functions, although this responsibility could be delegated as detailed below. The appropriate counterparts in other DOD components would have corresponding responsibilities.

The most important element is participation in the SEPR by peer software managers experienced in the application area, as well as by key stakeholders as either advocates or reviewers. Because the SEPRB's staffing from stakeholder organizations can vary considerably among systems, SEPRB representation is divided into mandatory and discretionary categories. The SAE must appoint representatives from mandatory stakeholders, but can include discretionary stakeholders as appropriate to the software engineering plan to be reviewed.

For systems subject to Milestone Decision Authority (MDA) at the service level or in the Office of the Secretary of Defense, the mandatory list of stakeholders and peer reviewers includes the following:

1. The PEOs (senior product-line officials) responsible for both development and post-deployment support for the candidate system and closely related systems;
2. Management and technical officials responsible for maintenance of the systems being specified or developed;
3. Representatives from user organizations, as appropriate;
4. Peer program managers with related software engineering management experience; and
5. Program managers for the system being specified, developed, or re-engineered (stakeholders, but reviewees rather than reviewers).

The discretionary list of stakeholders depends on the characteristics of the system being developed, but could include the following:

1. Program managers for development and support of critical related systems that must interoperate with or are otherwise closely affected by the system under review;
2. Representatives of the DOD community who have specific technical expertise and cognizance of emerging technologies;
3. Representatives of other program executive offices, program offices, or other components that are responsible for key common architectural frameworks; and
4. Representation, where appropriate, from the Office of the Secretary of Defense or the Joint Chiefs of Staff.

## Approval Authority and Milestones

For systems subject to MDA, approval authority for the process resides with the Assistant Secretary of Defense ($C^3I$) or the SAE, depending on the class of system. The direct management of the SEPR process would be carried out by the SAEs and their associated SEOs, with possible delegation to the PEO level. But the actual approval authority should not be delegated beyond the SAE. The Assistant Secretary of Defense ($C^3I$) would monitor the review process.

When significant deviations are needed from DOD's stated policy and principles, direct approval of the Assistant Secretary of Defense ($C^3I$) may be required; this should be determined when approval authority is delegated to the SAEs. It is the intent of these recommendations, however, that policy be framed with sufficient flexibility and outlook to the future that such deviations are not required in the ordinary conduct of business. It is also the intent that implementation be delegated to a level sufficient to ensure in-depth review of software engineering decisions. SEPRs are ongoing processes, with specific approvals pertinent to specific milestones. SEPRs must be linked, at a minimum, to DAB and MAISRC Milestones I and II.

Many smaller systems are subject to DOD software engineering policy, but not to MDA. For these systems, approval authority resides with the SAE, but there is flexibility with respect to delegation and the need for formal SEPRs. Normally, the SAE can delegate approval authority to a PEO or, for very small systems, to a major command. In the latter case, approval can be granted for a family of related small systems as a result of a software engineering plan for a single product line. But the committee suggests that for non-MDA systems, the decision to conduct a formal SEPR process (or some more expedient process) be required for warfighting software, and be a recommended practice at the discretion of the approval authority for other software.

Given that the Director of Defense Research and Engineering (DDR&E) is responsible for advanced (6.3a) research, the committee recommends that the DDR&E establish a software engineering review process that addresses issues pertinent to the efficient transition of software technologies associated with major 6.3a demonstration programs, including plans to modify prototype 6.3a software to conform to the committee's recommended policy on selection of programming language, as appropriate. The review criteria, which would be at the discretion of the DDR&E, would not need to use the SEPR process, thus enabling the DDR&E to manage the trade-off between efficient transitions, on the one hand, and responsiveness and flexibility of research programs to the emergence of new technologies and concepts, on the other.

## Submission of Software Engineering Plans

As envisioned by the committee, the SEPR process requires program managers responsible for MDA system specification, development, and major re-engineering efforts to submit a software engineering plan, preceded by a request to the SAE to convene a SEPRB. The SAE, considering the recommendations of the program manager and the cognizant PEO, would then select stakeholder organizations, which appoint representatives. For smaller systems, the SAE and PEO roles are further delegated, as indicated above.

It is the committee's intent that the approval authority would work with the SEPRB and the program manager to develop a software engineering plan suitable for the project and in conformance with all DOD policies. No entry into the DAB Milestone I and II reviews could be initiated without concurrence of the approval authority. Criteria used for evaluation of the software engineering plan should be defined by the approval authority.

The software engineering plan should be a simple document[3] and should cover areas relevant to the decision process, including the following:

- The system's scope and concept of operation;
- The key system and software requirements, including stakeholder needs;
- The key elements of the system and software architecture, including programming language decisions;
- The system and software life-cycle plans, including increments, budgets, and schedules; and
- A rationale demonstrating that the software can be developed within the budget and schedule specified in the life-cycle plan, can satisfy the requirements and key stakeholder needs, and can successfully support the concept of operation.

The SEPR approval authority, in consultation with PEO and program manager representatives, would develop specific criteria to be reviewed. The criteria for review could include, for example:

- System structural architecture, partitioning the system into components;
- Differentiation of key architecture requirements from secondary features and capabilities;
- Nature and extent of compliance of the architectural plan with related open-architecture and DOD framework common interfaces;
- Definition of increments and completion criteria (e.g., design to cost);
- Cost and risk management;
- Risk-management plan designed into early releases;
- Metrics for indicating progress and measuring completion of milestones;
- Major milestone content, evaluation criteria, and demonstration scenarios; and
- Basis for decisions to make, buy, or reuse components (see below).

For each subsystem or component in the system, the following areas should be addressed:

- Availability of COTS products, non-developmental items, and other existing or reusable components;
- Appropriateness of new development;
- Appropriateness of new development for reuse (capitalization) in related systems;
- Potential for reuse or insertion into other related systems—incentives can be established by additional resources provided by the PEO;
- Use of tooling and generators for development, and status of the tooling and generators;
- Degree of compliance with related interface or framework standards;
- Maintenance responsibility (government, contractor, or commercial); and
- Choice of programming language, subject to the recommended policy in Appendix A.

### Software Engineering Codes

As experience is gained, SAEs, PEOs, and other stakeholders will develop service-specific or domain-specific refinements of the review criteria listed in the previous section. For example, a service may designate conformance with a common architectural framework as a review item. These refinements may attain the status of software engineering "codes" (analogous to building codes) particular to a service or PEO product line. These would serve as "best practices" documents that would necessarily evolve over time, according to requirements and technology developments. They would also enable program managers to develop expectations concerning the SEPR process on the basis of their conformance with such codes.

## NOTES

1. This section and Appendix A present similar material in different formats. Appendix A was prepared by the committee to serve as a proposed revision to DOD Directive 3405.1 (DOD, 1987a); this section discusses the principles and rationale underlying the committee's suggested changes to that policy document.

2. The committee recommends using the term "software engineering plan reviews" rather than "architecture reviews" to emphasize the importance of integrating plans for products (i.e., architecture or building plan) with plans for process (e.g., increments, milestones, budgets).

3. The life-cycle objectives and life-cycle architecture milestones introduced in Boehm (1996) provide guidelines for the level of detail of a software engineering plan desired at DAB and MAISRC Milestones I and II.

# 5

# Implementation of Recommended Strategy for Investment in Ada

This chapter describes the committee's recommended $15 million annual investment strategy for Ada in the context of the focus on use of Ada for DOD's warfighting software. The components of the strategy are maintaining and enhancing the Ada 95 language; ensuring support for Ada's compilers, tools, and application programming interfaces (APIs) for warfighting systems; stimulating Ada education and curriculum development; and providing support for a centralized source of Ada information and expertise within DOD. The chapter concludes by providing a detailed plan and funding breakdown for the recommended investments in Ada.

## GOALS OF THE INVESTMENT STRATEGY

It is important to state explicitly what the goals of the proposed investment and procurement strategy for Ada are, and what they are not. Using Ada benefits DOD mainly because of the language's orientation toward reliability. Ada's features are designed to maximize the chance that a coding error will be detected at compile time, and failing that, at run time. As illustrated in "Technical Evaluation of Ada 95 and Other Third-Generation Programming Languages" in Chapter 2, Ada is ahead of languages such as C and C++ in that it provides strong support for compile-time and run-time consistency checking. Although other languages may approach Ada in their ability to check reliability and consistency, no other third-generation programming language (3GL) has achieved comparably widespread use for critical (high-assurance, and often real-time) systems.

The first goal is to *ensure that Ada tools for critical warfighting software development are superior to those in other languages.* This is at the heart of the strategy to keep Ada competitively superior for the development of warfighting software systems. It involves a proactive strategy of seeking out and expediting (via complementary funding) development of such tools for the commercial marketplace.

The second goal is to *ensure that robust Ada compilers and associated tools for host and target computers used in DOD's warfighting systems remain available and reasonable in cost.* It is not a goal to make Ada compilers as inexpensive or ubiquitous as C compilers. The strategy recognizes that a compiler for a reliability-oriented language like Ada, with large numbers of compile-time and run-time consistency checks, will inevitably be larger and more expensive to develop and maintain than other 3GL compilers. A reasonable target would be for a production-quality Ada compiler to be available at a price comparable to the cost of a production quality C compiler, plus additional production quality third-party compile-time and run-time checking tools for C (for example, a "lint"-like tool and a "purify"-like tool). The strategy also recognizes that any compiler for an embedded system with associated robust cross-"debugging" support and integration with an appropriate real-time executive will be more expensive than a personal computer (PC) native compiler. It is not expected that Ada cross-compilers will ever match the price of off-the-shelf PC native compilers.

For similar reasons, it is not a goal that Ada become the predominant programming language for commercial programming. For critical systems programming, Ada already has a significant market share. Nevertheless, this is a relatively small marketplace, and DOD is a major player, meaning that it will have to continue to invest in the market for tools that support the development of critical systems.

The third goal is to *ensure that an Ada-compatible interface is readily available for all off-the-shelf software components used in DOD's warfighting systems.* Such systems include relevant database management systems, operating systems, real-time executives, and networking packages. It is not a goal that such components necessarily be written in Ada.

The fourth goal is that *more educational institutions provide exposure to Ada concepts in the context of a software engineering curriculum, particularly in courses covering critical systems development.* It is not a goal that Ada become the predominant teaching language (though Ada has advantages for teaching because of its readability and consistency checking). Rather, the goal is to increase the exposure to and use of Ada in the educational environment, thereby increasing the pool of programmers familiar with Ada and reducing the ultimate training costs for DOD's contractor community.

A related goal is to increase the use of Ada in software engineering research, which will enhance the Ada technology base for warfighting systems, as well as increase exposure to and understanding of Ada in the academic and research communities. It is not suggested that all DOD-sponsored research use Ada or that all DOD research prototypes be written in Ada. Rather, it is recommended that critical-systems software research emphasize Ada's use, that more research programs include Ada in the set of languages they consider, and that an increased number of advanced technology tools support Ada in addition to other languages that are supported.

The final goal relates to *centralized support and infrastructure for DOD's use of Ada.* DOD can continue to benefit from using Ada for its critical systems, but the benefits will be offset by other cost increases if each project needs to maintain its own support infrastructure for Ada. By continuing to support a centralized organization like the Ada Joint Program Office, even if its mission is redirected toward ongoing support rather than product development, DOD can achieve economies of scale. It is not suggested that such an organization directly support the development of new Ada compilers. Rather, it is emphasized that a centralized organization is necessary to provide expertise, resource directories (such as the World Wide Web site supported by the Ada Joint Program Office), technology management, and technology transition.

## ADA INVESTMENT STRATEGY

DOD can benefit from providing ongoing support for Ada technology by building on its significant investment in Ada 83, and the Ada 95 revision. In addition, DOD should use its leverage as a large customer for commercial hardware and software products to expand the availability of commercial tools and components that support Ada, which would help the Ada market become more self-sustaining.

Continued investment by DOD is required in the following areas:

1.  Supporting the continued maintenance of the Ada 95 language, and the enhancement of Ada-related secondary standards in areas relevant to defense systems;

2.  Ensuring that, for all DOD warfighting systems, hardware has robust Ada compiler and tool support, commercial off-the-shelf (COTS) software has Ada-compatible APIs, and Ada tools for critical warfighting software are superior to those in other languages;

3.  Providing support for the continued development of curricula related to Ada and software engineering in the nation's colleges and universities, particularly in the area of critical systems; and

4.  Funding a centralized DOD source of information and expertise on Ada and related software engineering technologies.

Each area is discussed in more detail below.


## Language Maintenance and Enhancement

The Ada 95 standard requires ongoing support.  Parts of the standard require interpretation, implying the need for a group of experts to meet periodically to resolve these issues.  Any large software system will develop unanticipated requirements for improvement as its use becomes widespread, and Ada 95 is no exception.  In addition, advanced areas such as security, distribution, or persistence require ongoing development of "secondary" standards and accompanying validation test suites to systematize the best way to use Ada in these areas.  For Ada, there is an International Organization for Standardization working group (ISO WG9), which has a number of "rapporteur" groups that coordinate these maintenance and secondary standardization tasks.  Funding of these groups is essential to make their activities effective and timely, and DOD must provide some of the critical funding to keep these groups active.  Even in the area of compiler validation, where the Commerce Department's National Institute of Standards and Technology is now providing a nominal level of support, DOD may find it necessary to provide funding to ensure that its high-priority needs are met in a timely manner.


## Support for Ada Compilers, Tools, and Application Programming Interfaces

Ada 95 is a technically successful revision of Ada 83.  Like all standards, however, it needs "champions" for its adoption, since there are always costs involved in adopting a standard or revision. Now that the Ada 95 revision is complete, the user and supplier communities must be given incentives to adopt the standard.  Although DOD is no longer a dominant player in the software market, it is still a large customer, and arguably the primary customer for critical systems.  Any major customer can establish incentives to motivate its suppliers to support a preferred standard; in order for DOD to play this role for its critical warfighting systems, it must emphasize the importance of providing support for Ada compilers and tools, and for Ada APIs for warfighting software applications.  DOD should similarly use its market power to make it possible to tailor a warfighting system to operate efficiently in Ada, even if the original system was built largely using COTS components or non-Ada technologies.  This means that DOD should require that all components included in DOD warfighting software have a demonstrable, full-function, Ada-callable interface.  For example, all operating systems used in DOD warfighting systems should include an Ada-callable interface, sufficient for exercising all capabilities of the operating system necessary for warfighting systems.  The Ada-callable interface must be kept up to date by the operating system supplier; in order for a capability to be considered acceptable for delivery, it must be accessible via Ada. Similarly, a hardware device that is controlled by software should be delivered with a demonstrable Ada-callable interface before it is accepted for a DOD warfighting system.

Another incentive to encourage use of Ada is to ensure that all computers used for developing DOD warfighting software include an Ada compiler at a price and quality comparable to those of other compilers. The criteria for choosing a computer for hosting software development should include the availability, price, and quality of the Ada compiler. This will help to ensure that hardware vendors wishing to sell in the DOD warfighting-systems marketplace make certain that there is a high-quality, reasonably priced Ada compiler on their platform. Warfighting systems procurements should stipulate that Ada compilers be available at the same time the equipment is available, rather than months or years later. This approach will ensure that Ada development remains viable on hardware platforms relevant to DOD.

For Ada to be a viable option for a given system, appropriate Ada compilers and associated tools must be available for the relevant host and target hardware. In the mid 1980s, most hardware vendors made an effort to ensure that an Ada compiler and associated tools existed for their hardware. Recently, however, the burden seems to have shifted from the hardware vendors to the individual system-development contractors. The linkage between purchasing hardware and compiler availability seems to have been lost.

In addition to robust support for compilers, Ada-compatible APIs to relevant COTS software components are also important to make the use of Ada cost-effective. As in the hardware area, the linkage between selling software components to DOD and providing appropriate Ada support seems to have been lost. For example, rather than requiring that database management systems for warfighting software systems come with an Ada-compatible interface, the burden seems to have shifted to individual contractors to acquire or develop such interfaces. Because COTS software products are being continually enhanced, this approach is even less satisfactory than in the hardware area.

Providing multilingual interfaces to COTS software products is now done as a matter of course. As with hardware, software vendors who wish to compete for DOD warfighting systems business should recognize that including Ada among the languages supported is a normal cost of doing business. The cost of including an Ada-compatible interface is relatively small. With the advent of multilingual interface standards such as CORBA IDL,[1] this proposed requirement becomes even more practical. On the other hand, when the burden of providing an Ada-compatible interface is left to individual contractors, the cost is greater, particularly given the need to keep up with ongoing enhancements.

DOD should re-create this linkage between robust Ada compiler and tool support and the choice of hardware. Hardware vendors who wish to compete for DOD warfighting systems procurements should see the provision of appropriate Ada support as a requirement, in the same way that hardware vendors have recognized the need to provide a POSIX-compliant[2] operating system to satisfy DOD's open systems requirements.

As defined in Chapter 2, "warfighting systems" are the high-assurance, performance-critical portions of systems for weapon control, electronic warfare, wideband real-time surveillance, real-time battle management, and special battlefield communications. The Ada support requirements recommended above apply only to hardware and software vendors choosing to support such systems. Examples of Ada tools that support critical warfighting software engineering include Ada-oriented tools for real-time and fault-tolerant software; tools for software reliability, availability, survivability, security, and safety assurance; tools to facilitate effective use of Ada 95's new, object-oriented, real-time, and distributed-system capabilities for warfighting systems; and Ada-oriented tools for special warfighting applications domains (e.g., real-time distributed avionics systems test, debugging, and performance tuning).

## Curriculum Development

For the past several years, the Ada Joint Program Office has provided support for the development of Ada-based curriculums at a number of U.S. colleges and universities.  This has contributed to the growing number of colleges and universities that include Ada and software engineering in their computer science programs.  As discussed in Chapter 1, the number of courses in Ada and in software engineering has increased recently, but there are still many colleges and universities that provide little or no exposure to the kinds of software challenges faced in the development of complex critical systems.  One of the frequent concerns of defense contractors has been the availability of well-trained software engineers.  Since DOD's investments in Ada curriculum development seem to have assisted in such training, the grants for this purpose should continue.  An effort to distribute the grants geographically is advisable, to ensure that Ada instruction is available in areas where DOD warfighting systems are developed.

In addition to supporting Ada-related curriculum development, DOD should ensure that some portion of its software research budget is focused on enhancing the Ada technology base.  A growing number of tools and techniques are being investigated in the research community for enhancing the productivity and quality of software development.  Whether these tools are actually written in Ada is not particularly relevant to DOD.  What is relevant, however, is whether these tools can be used to support systems with components written in Ada.  With the availability of the GNAT compiler in source code (for details, see Chapter 1), there are fewer obstacles to the use of Ada as a vehicle for research in programming languages and programming support tools.

In keeping with the competitive strategy recommended in earlier chapters, this investment package should place a high priority on Ada-based research and curriculum development for software needs in warfighting application domains (e.g., for real-time, high-assurance software).

## Centralized Support Organization

In testimony provided to the committee, the importance of a centralized organization to support Ada and related software engineering technologies was repeatedly stressed.  The Ada Joint Program Office has served this function over the past decade.  While it is not the role of this committee to recommend the disposition of this function, it may be appropriate to consider transferring executive responsibility to a higher-level organization with cognizance over warfighting software, such as the Office of the Assistant Secretary of Defense ($C^3I$) or the Undersecretary of Defense (Acquisition and Technology).  With regard to choosing a support organization, it is worth noting that Ada is one of several software technologies important to defense systems.  Thus, combining support for Ada with support for other important software technologies might save resources and provide better service overall.

## DETAILED PLAN FOR INVESTMENTS IN ADA TECHNOLOGY AND SUPPORT

Chapter 3 presented the argument that if Ada 95 is well supported, it could create a significant warfighting competitive advantage for DOD.  The committee recommends the following investment strategy to help bring this about.  The overall $15 million annual budget needed for ongoing enhancement of Ada-related technology, and ongoing support for the effective use of Ada within the DOD community, can be divided into two major areas.  The first is for funding contracts with organizations outside DOD.  The second area is for funding the ongoing activities of a centralized support organization within DOD oriented toward programming languages.

The total annual budget of $11.5 million for contracts should be allocated as follows:

- Education contracts for Ada and software engineering curriculum development:
  $1.5 million, based on approximately 20 awards at $75,000 each;
- Application programming interface binding development and maintenance:
  $1.0 million, based on approximately 5 awards at $200,000 each;
- Expert consultation to support language maintenance and evolution:
  $0.6 million, based on approximately 15 part-time experts at $40,000 each;
- Maintenance and enhancement of validation suites:
  $0.4 million, based on approximately 2 awards at $200,000 each; and
- Development of warfighting-critical tools and technology:
  $8.0 million, based on approximately 15 fully funded academic/research-oriented awards averaging $200,000 each, and 20 co-funded industry/development-oriented awards averaging $250,000 in DOD funds each.

The annual budget of $3.5 million for a centralized support organization should be distributed as follows:

- Compliance monitoring to identify gaps in technology that are hindering compliance and to identify possible remedies:
  $0.5 million;
- Support of an information clearinghouse and a World Wide Web home page:
  $1.0 million;
- Development of DOD-oriented education and training materials, Ada 83/Ada 95 transition support and tools, and technology insertion:
  $1.0 million; and
- Metrics on programming language usage, cost-effectiveness, performance, and defect rates; collection, analysis, and dispersal of the information:
  $1.0 million.

As stated above, it is important to emphasize that these are not one-time investments, but rather continuing investments to maintain and enhance the effectiveness of Ada for DOD. It is also important to recognize that this $15 million annual DOD investment is designed to expedite desired capabilities by leveraging the existing $200 million annual Ada market. It does not represent the entire investment required for Ada support.

## CONCLUSION

If DOD continues to specify Ada as the preferred programming language—for any application domains—it is essential that it make investments such as those outlined in this chapter. The only other sources of Ada support are individual projects, and burdening them would be inefficient, duplicative, and a major disincentive for projects to use Ada. From a vendor's standpoint, it would put Ada into the same narrow niche as Jovial, which has very little support for compilers, tools, bindings, and run-time applications.

Furthermore, failure to make the relatively modest level of investment described above would imperil the existing body of Ada code for DOD's weapon systems. Independent of the question of future systems development, the legacy of developed Ada code requires a robust development community. The economic analysis presented in Chapter 3 shows that the cost savings from maintaining this legacy code alone are sufficient to justify the recommended $15 million annual investment in Ada.  But even further, as discussed in the business-case analysis, the investment provides DOD with a key competitive advantage in warfighting software to more completely achieve its objective of military information dominance.

## NOTES

1. Common object request broker architecture interface definition language.
2. Portable Open Systems Interface.

# Bibliography

Aerospace Industries Association (AIA). 1996. "AIA Position Statement on Ada." Unpublished manuscript. AIA, Washington, D.C., July 12.

Aley, James. 1996. "Give It Away and Get Rich." *Fortune*, June 10, p. 98.

AT&T. 1993. "Best Current Practices: Software Architecture Validation." AT&T, Murray Hill, N.J.

Boehm, Barry. 1981. *Software Engineering Economics*. Prentice-Hall, Englewood Cliffs, N.J.

Boehm, Barry, and W.E. Royce. 1988. "TRW IOC Ada COCOMO: Definition and Refinements." *Proceedings of the 4th COCOMO Users Group*, Pittsburgh, Pa., November.

Boehm, Barry. 1996. "Anchoring the Software Process." *IEEE Software*, Vol. 13, No. 4, July, pp. 73-82.

Boehm, Barry, et al. 1996. "The COCOMO 2.0 Software Cost Estimation Model: A Status Report." *American Programmer*, Vol. 9, No. 7, July, pp. 2-17.

Booch, Grady. 1987. *Software Engineering with Ada,* Third Edition, Chapter 3: "The History of Ada's Development." Benjamin/Cummings, Menlo Park, Calif.

Brooks, Jr., Frederick P. 1986. "No Silver Bullet—Essence and Accidents of Software Engineering." *Information Processing* 86, pp. 1069-1076. (Reprinted in *IEEE Computer Magazine*, April 1987.)

Brown, Norm. 1996. "Industrial Strength Management Strategies." *IEEE Software*, Vol. 13, No. 4, July, pp. 94-103.

Carleton, Anita D., et al. 1992. "Defining and Using Software Measures." Software Engineering Institute Technical Reports, CMU-SEI-TR-11, 19-23. Software Engineering Institute, Pittsburgh, Pa.

Computer Science and Telecommunications Board (CSTB). 1989. *Scaling Up: A Research Agenda for Software Engineering*. National Academy Press, Washington, D.C.

CTA. 1991. "Survey and Analysis of Productivity and Cost Data on Ada and C++ Software Development Programs." CTA Incorporated, Burlington, Mass., June 26.

Department of Defense (DOD). 1976. Directive 5000.29, "Management of Computer Resources in Major Defense Systems." April 26.

Department of Defense (DOD). 1987a. Directive 3405.1, "Computer Programming Language Policy." April 2.

Department of Defense (DOD). 1987b. "Military Software." Defense Science Board Task Force on Military Software (Frederick P. Brooks, Jr., Chair), Office of the Under Secretary of Defense for Acquisition, DOD, Washington, D.C., September.

Department of Defense (DOD). 1991. "Draft DOD Software Technology Strategy." Director of Defense Research and Engineering, DOD, Washington, D.C., December.

Department of Defense (DOD). 1992. "Delegations of Authority and Clarifying Guidance on Waivers from the Use of the Ada Programming Language." April 17.

Department of Defense (DOD). 1994a. "Report of the Defense Science Board Task Force on Acquiring Defense Software Commercially." DOD, Washington, D.C., June.

Department of Defense (DOD). 1994b. "Specifications and Standards: A New Way of Doing Business." June 29.

Department of Defense (DOD). 1995a. "New World Vistas: Air and Space Power for the 21st Century. Information Technology Volume." Air Force Scientific Advisory Board, Washington D.C.

Department of Defense (DOD). 1995b. "Dual Use Technology: A Defense Strategy for Affordable, Leading-Edge Technology." Under Secretary of Defense for Acquisition and Technology, DOD, Washington, D.C., February.

Department of Defense (DOD). 1995c. "ARPA Software Review Panel—Final Report." DOD, Washington, D.C., October.

Department of Defense (DOD). 1996a. "Practical Software Measurement Version 2.1." Joint Logistics Commanders, Joint Group on Systems Engineering, DOD, Washington, D.C., March 27.

Department of Defense (DOD). 1996b. Directive 5000.1, "Defense Acquisition." February 21.

Department of Defense (DOD). 1996c. Regulation 5000.2-R, "Mandatory Procedures for Major Defense Acquisition Programs (MDAPs) and Major Automated Information System (MAIS) Acquisition Programs." March 15.

Department of Defense (DOD). 1996d. "Joint Warfighting Science and Technology Plan." Office of the Secretary of Defense, DOD, Washington, D.C., May.

Druffel, Larry. 1993. "Ada Position Paper." Unpublished manuscript, December 1.

Emery, James C., and Martin J. McCaffrey. 1991. "Ada and Management Information Systems: Policy Issues Concerning Programming Language Options for the Department of Defense." Naval Postgraduate School, Monterey, Calif., June.

Emery, James C., and Dani Zweig. 1993. "The Use of Ada for the Implementation of Automated Information Systems Within the Department of Defense." Naval Postgraduate School, Monterey, Calif., December 28.

Feigenbaum, Edward. 1996. "The Ada Mandate." Unpublished manuscript, Chief Scientist, U.S. Air Force.

Feldman, Michael B. 1996. "Ada as a Foundation Programming Language." Available on-line at http://www.seas.gwu.edu/faculty/mfeldman/CS1-2.html#2.

Fisher, D.A. 1976. "A Common Programming Language for the Department of Defense—Background and Technical Requirements." Report P-1991. Institute for Defense Analyses, p. 6. (Cited by G. Booch in Software Engineering with Ada, Third Edition, 1987, p. 9.)

Fisher, David. 1974. "Automatic Data Processing Costs in the Defense Department." IDA Paper P-1046. Institute for Defense Analyses, Alexandria, Va., October.

Frazier, Thomas P., and John W. Bailey. 1996. "The Costs and Benefits of Domain-Oriented Software Reuse: Evidence from the STARS Demonstration Projects." IDA Paper P-3191. Institute for Defense Analyses, Alexandria, Va., June.

General Accounting Office. 1989. "Programming Language: Status, Costs, and Issues Associated with Defense's Implementation of Ada." GAO/IMTEC-89-9. General Accounting Office, Washington, D.C., March.

General Accounting Office. 1991. "Programming Language: Defense Policies and Plans for Implementing Ada." GAO/IMTEC-91-70BR. General Accounting Office, Washington, D.C., September.

General Accounting Office. 1993. "Software Reuse: Major Issues Need to Be Resolved Before Benefits Can Be Achieved." GAO/IMTEC-93-16. General Accounting Office, Washington, D.C., January.

Giallombardo, Robert J. 1992. "Effort and Schedule Estimating Models for Ada Software Development." MTR 11303. MITRE Corporation, Bedford, Mass., May.

Hook, Audrey A., et al. 1991. "Availability of Ada and C++ Compilers, Tools, Education and Training." IDA Paper P-2601. Institute for Defense Analyses, Alexandria, Va., June 12.

Hook, Audrey A., et al. 1995. "A Survey of Computer Programming Languages Currently Used in the Department of Defense." IDA Paper P-3054. Institute for Defense Analyses, Alexandria, Va., January.

Horowitz, Barry. 1991. "The Importance of Architecture in DOD Software." M91-35. MITRE Corporation, Bedford, Mass., July.

IBM Federal Systems Division. 1985. "Language Selection Analysis Report." FAA-85-S-0874. Prepared for the Federal Aviation Administration, Gaithersburg, Md., May.

IIT Research Institute (IITRI). 1996. "Catalog of Resources for Education in Ada and Software Engineering, Version 8.0." Prepared for the Ada Joint Program Office, Arlington, Va., January.

Jensen, L.W., and S. Lucas. 1983. "Sensitivity Analysis of the Jensen Software Model." Hughes Aircraft Co., Los Angeles, Calif.

Jones, Capers. 1994. "The Economics of Object Oriented Software." *American Programmer*, Vol. 7, No. 10, October, pp. 28-35.

Jones, Capers. 1995. "Backfiring: Converting Lines of Code to Function Points." *Computer*, Vol. 28, No. 11, November, pp. 87-88.

Jones, Capers. 1996a. "Estimating and Measuring Object-Oriented Software." Unpublished manuscript, February 1.

Jones, Capers. 1996b. "The Economic Impact of the Year 2000 Software Problem in the United States—Version 2." Software Productivity Research Inc., Burlington, Mass., February 22.

Jones, Capers. 1996c. "Programming Languages Table, Version 8.2." Software Productivity Research Inc., available on-line at http://www.spr.com/library/langtbl.html.

Jones, Capers. 1996d. "Using Function Points to Evaluate CASE Tools, Version 4." Software Productivity Research Inc., Burlington, Mass., August 6.

Landry, Huet C. 1996. "Comments on Ada Mandate." Unpublished manuscript, Defense Information Systems Agency/JEBEB, Software Engineering Standards.

Lawlis, Patricia K. 1996. "Guidelines for Choosing a Computer Language: Support for the Visionary Organization." C.J. Kemp Systems Inc., March.

Lubashevsky. 1996. "'Backfire' Will BACKFIRE." *Measure Up.* Software Productivity Solutions Inc., Indiatlantic, Fla., January, pp. 1-6.

Maranzano, Joe. 1995. "System Architecture Validation Review Findings." AT&T Software Technology Center, April.

Masters, Michael W. 1996. "Programming Languages and Life-Cycle Cost." Naval Surface Weapons Center, Dahlgren, Va., March 18.

McGarry, Frank, et al., 1994. "Software Process Improvement in the NASA Software Engineering Laboratory." CMU SEI-94-TR-22. Software Engineering Institute, Pittsburgh, Pa., December.

Mosemann, Lloyd K. 1991. "Ada and C++: A Business Case Analysis." Deputy Assistant Secretary of the Air Force, Washington, D.C., July 9.

National Academy of Engineering. 1996. *Defense Software Research, Development and Demonstration: Capitalizing on Continued Growth in Private-Sector Investment.* National Academy Press, Washington, D.C., March.

National Research Council. 1995. *Research-Doctorate Programs in the United States: Continuity and Change.* National Academy Press, Washington, D.C.

Paulk, Mark C., et al. 1993. "Capability Maturity Model for Software, Version 1.1." CMU/SEI-93-TR-24. Software Engineering Institute, Carnegie Mellon University, Pittsburgh, Pa., February.

Perry, William. 1996a. "National Academy of Engineering, Bueche Prize Acceptance Address." Transcript. National Academy of Engineering, Washington, D.C., October 2.

Perry, William. 1996b. "Defense in an Age of Hope." *Foreign Affairs*, November–December, pp. 64-79.

Porter, Michael E. 1990. *The Competitive Advantage of Nations.* The Free Press, New York.

Powell, Colin. 1992. "C$^4$I for the Warrior." DOD Joint Chiefs of Staff, Washington, D.C., June 12.

Quade, E.S., ed. 1964. *Analysis for Military Decisions.* Rand McNally, Chicago.

Quade, E.S., and Boucher, W.I., eds. 1968. *Systems Analysis and Policy Planning: Applications in Defense.* American Elsevier, New York.

Reifer, Donald J. 1996. "Quantifying the Debate: Ada Versus C++." *Crosstalk: The Journal of Defense Software Engineering.* Hill AFB, Utah, July.

Riehle, Richard. 1996. "Ada: An Update." *Object Magazine*, June, pp. 50-52.

Royce, Walker. 1990. "TRW's Ada Process Model for Incremental Development of Large Software Systems." *Proceedings ICSE 12*, IEEE/ACM, March, pp. 2-11.

Shaw, Mary, and David Garlan. 1996. *Software Architecture: Perspectives on an Emerging Discipline.* Prentice-Hall, Englewood Cliffs, N.J.

Telos Corporation. 1994. "Ada Marketing Communications Study Precampaign Survey." Telos Corporation, May.

Tokar, Joyce. 1996. "Ada 95: The Language for the '90s and Beyond." *Object Magazine*, June, pp. 53-56.

TRW.  1991.  "Ada and C++:  A Lifecycle Cost Analysis."  TRW Inc., Redondo Beach, Calif., June 1.

TRW.  1991.  "Case Study:  Ada and C++ Cost Comparison for CCPDS-R."  TRW Inc., Redondo Beach, Calif., June 1.

U.S. Army. 1992. "Implementation of the Ada Programming Language." HQDA LTR 25-92-1. September 18.

U.S. Army. 1994. "Change to HQDA Letter 25-92-1, Implementation of the Ada Programming Language."  HQDA LTR 25-94-1.  July 17.

U.S. Army. 1995. "Change to HQDA Letter 25-94-1, Implementation of the Ada Programming Language."  HQDA LTR 25-95-1.  July 17.

U.S. Navy. 1994.  "Ada Programming Language Policy,"  SECNAVINST 5234.2A.  April 28.

U.S. Navy. 1994.  "Ada Implementation Guide," 2 Vol. Naval Information Systems Management Center, April.

Vyssotsky, Victor.  1996.  "Whither Ada?"  Unpublished manuscript,  May 12.

Waligora, Sharon, John Baily, and Mike Stark.  1995.  "Impact of Ada and Object-Oriented Design in the Flight Dynamics Division at Goddard Space Flight Center."  NASA-SEL Report, SEL-95-001.  Goddard Space Flight Center, March.

Weiderman, Nelson.  1991.  "A Comparison of Ada 83 and C++."  SEI-91-SR-4.  Software Engineering Institute, Pittsburgh, Pa., June.

Zeigler, Stephen F.  1995.  "Comparing Development Costs of C and Ada."  Rational Software Corporation, Santa Clara, Calif., March 30.

# Appendixes

# Appendix A

# DOD Draft Software Management Policy Directive with Further Modifications Suggested by the Committee

## INTRODUCTION

The Committee on the Review of the Past and Present Contexts for the Use of Ada in the Department of Defense reviewed the DOD policy currently in use for programming language selection ("Computer Programming Language Policy," DOD Directive 3405.1, dated April 2, 1987), as well as two different draft revisions of that policy. This appendix contains the most recent draft (dated May 15, 1996) reviewed by the committee and incorporates modifications suggested by the committee to make the directive consistent with the recommendations presented in the main text of this report. Modifications are noted in italic font.

This modified draft directive is intended to serve as a "template" for development of the new DOD Directive 3405.1. The "enclosures" that are typically attached to DOD directives have been omitted for brevity. However, a list of references follows the text of the draft directive, and technical terms are defined in Appendix C of this report; both sets of documentation are suitable as enclosures for the revised formal DOD directive. The enclosure titled "Ada Waiver Procedures" has been omitted from the template because the committee eliminated the waiver process in its recommended policy. Other emendations of the draft text to condense wording or otherwise revise original text are not strictly documented; comparison with the May 15, 1996, draft directive shows minor changes not accommodated by the device of italicizing more substantial revisions.

## PROPOSED TEMPLATE FOR DOD DIRECTIVE ON SOFTWARE MANAGEMENT

A.      REISSUANCE AND PURPOSE

This Directive:

1.  Updates and establishes policy for management of software developed, used, or maintained by, or for, the Department of Defense (DOD).

*2.  Is used in software management decisions across a functional or mission area, domain, or product line. It contains broad software engineering and programming language policy that will be followed by DOD.*

*3.  Establishes the requirement for a Software Engineering Plan Review Board (SEPRB) by the Office of the Secretary of Defense (OSD), the Military Departments (including the National Guard and Reserve Components), and the DOD components.*

4.  Supersedes reference (a); cancels references (b) and (c); implements Federal Information Resource Management Regulation (FIRMR) Subpart 201-24.201, Federal Software Exchange Program (reference (d)); and supports DOD Directive 8000.1 (reference (e)) and DOD Directive 5000.1 (reference (f)) and DOD Instruction 5000.2 (reference (g)).

5.  Authorizes publication of DOD Instruction 3405.1, "Software Management Implementation."

## B.        APPLICABILITY AND SCOPE

This Directive applies to:

1.  The Office of the Secretary of Defense (OSD), the Military Departments (including the National Guard and Reserve components), the Chairman of the Joint Chiefs of Staff and the Joint Staff, the Unified Combatant Commands, the Inspector General of the Department of Defense, the Defense Agencies, and the DOD Field Activities (hereafter referred to collectively as "the DOD Components").

2.  All software developed, acquired, or used by the DOD, including that managed in accordance with DOD Directive 5000.1 (reference (f)).

3.  DOD research and development activities funded by *6.4 and 6.5* appropriations as defined in Volume 2, DOD 7000.14-R, (reference (l)).

*4.  Software developed, acquired, or used by DOD research and development activities, and funded by 6.1, 6.2, and 6.3a appropriations, is exempted from this directive.*

## C.        DEFINITIONS

*Committee note: Terms used in this directive and modifications to it are defined in Appendix C of this report, and those definitions are suitable for inclusion as an enclosure in a new DOD directive.*

## D.        POLICY

It is DOD policy to:

*1.  Perform trade-off and business-case analysis in the development and acquisition of affordable, rapidly produced, high-quality software. Quality includes functionality, fitness for a purpose, assurance (i.e., reliability, survivability, availability, safety, security), efficiency, ease of use, interoperability, future adaptability (i.e., extensibility, maintainability, portability, and compliance with standards), and the development of DOD software expertise. Cost includes full life-cycle cost, consequence of system failure, impact on system operational costs, and use of other scarce resources such as expert personnel.*

*2.  Exploit and contribute to open standards-based technical architectures that support rapid, flexible, and incremental software improvements, and that accommodate increasing reliance on the commercial sector to satisfy evolving mission and functional requirements. Exploit and contribute to software architectures that serve as the basis for management and investment decisions for reuse opportunities, interoperability requirements, and development or product lines and product-line components. Facilitate the reuse of software assets.*

3. Define mission and functional requirements so that commercial and non-developmental items may be used to fulfill such requirements. To the maximum extent practicable, modify requirements and conduct market research and analysis, prior to commencing a development effort, to take advantage of a commercial or non-developmental "best value" solution. Give preference to commercial off-the-shelf (COTS) items first and non-developmental items (NDI) second when satisfying software asset requirements.

4. Implement and continuously improve software process management and software engineering disciplines. Use software engineering environments that facilitate software process management and software engineering disciplines.

5. Employ software developers who possess mature software engineering capabilities. Software developers should have a successful past performance record, experience in the software domain or product line, a mature software development process, and evidence of use and adequate staffing and training in software methodologies, tools, and environments.

6. Use metrics when monitoring and managing production and delivery of software assets, evaluating maintenance and management practices, implementing software architectures and product lines, *and effecting continuous software process improvement.*

7. Enforce compliance with contractual terms and conditions for use of software, including copyright and license agreements. *Centralize this function to the maximum extent practicable.*

8. *Use commercial fourth-generation programming languages (4GLs) where appropriate, when they provide significant improvements in productivity, usability, maintainability, and portability. Selection of 4GLs and associated tools must be based on established acceptance in the commercial marketplace where benefits of the technology have been demonstrated. Any software code generated using a 4GL must also be maintained in the 4GL.*

9. *Use the highest-level language meeting quality, cost, and scheduling constraints for each software component. Principles for choice of this language are as follows:*

a. *Higher-level languages are generally preferable to lower-level languages.*

b. *Standardized and non-proprietary languages are preferred. Using standards increases portability of code and programmers. Non-proprietary languages reduce the risk of vendor lock-in.*

c. *New languages must not be developed as part of a system development, except for domain-specific languages providing directives for application generators.*

d. *Quality, time, and cost factors should be considered in selecting a language.*

10. *Use the Ada programming language (reference (i)) to develop software subsystems when* **all** *of the following apply:*

a. *The application is in a warfighting application area (i.e., weapon control, electronic warfare, wideband real-time surveillance, battle management, special battlefield communications).*

b. *Maintenance will be government-directed.*

c. *The expected size of the subsystem exceeds 10,000 lines of code, or the subsystem is critical.*

d. *There is no better COTS, NDI, 4GL, or higher-level solution consistent with quality and cost goals.*

e. *There is no life-cycle cost-effectiveness justification for using another programming language.*

f. *The code is new or re-engineered.*

*In cases meeting the above criteria, the required compliance level is set at 95% of the source lines of code.  Up to 5% of the code can be written in other languages to facilitate component integration.*

11.  *When the application is in a non-warfighting application area (e.g., office and management support, personnel, logistics, medicine, routine operations support), Ada will be analyzed as an option when substantial 3GL development is to be performed.  The analysis will be in accordance with the principles set forth in D.9 above.*

E.    RESPONSIBILITIES

1.    The Under Secretary of Defense (Policy) shall ensure that requests from DOD Components for guidance on international transfer or export of DOD software are processed and appropriate guidance on such release is provided.

2.    The Office of the Assistant Secretary of Defense (OASD) for $C^3I$ shall:

a.  Provide policy, guidance, and oversight for the management of software consistent with applicable directives, and may issue additional instructions related to implementation of this Directive.

b.  Issue policies and guidance to implement DOD software reuse practices and the Federal Software Exchange Program (reference (d)).

c.  Direct DOD Components to establish programs, as appropriate, to enhance the software engineering processes and the transition of technologies from commercial and research programs into applications within weapon, automated information systems (AISs), and command and control systems.

d.  *Establish an OSD-level Software Engineering Plan Review (SEPR) process that will review all software architecture plans for any acquisition subject to OSD milestone decision authority (MDA).  The purpose of this review will be to approve and certify the software engineering plans for the system software prior to Milestone I and II reviews.  Certification indicates that the software plan conforms to the policy and principles contained within this Directive.  For these reviews, the acquisition program shall establish a program-specific Software Engineering Plan Review Board (SEPRB).  The SEPRB will be composed of at least 5 members who are software experts, will include key system stakeholders (e.g., users, maintainers, interoperation experts), and will be chaired by a representative of the OASD ($C^3I$).*

e.  *Establish a process for periodic review of SEPRB reviews performed by DOD components.*

f.  Identify research and development requirements to the Director, Defense Research and Engineering, for inclusion in research and development programs.

3.    The Head of Each DOD Component shall:

a.  Initiate appropriate strategies and actions to implement the policies in Section D within their areas of responsibility.

b.  *Establish  a component-level SEPR process that will review all software architecture plans for any acquisition subject to component MDA.  The purpose of this review will be to approve and certify the plans for the system software prior to Milestone I and II reviews.  Certification indicates that the software plan conforms to the policy and principles contained within this Directive and applicable component policies.  For these reviews, the acquisition program shall establish a program-specific SEPRB.  The SEPRB will be composed of at least 5 members who are software experts, will include key system stakeholders (e.g., users, maintainers, interoperation experts, program executive officials), and will be chaired by a member appointed by the service or component acquisition executive.*

c. *Establish and monitor a SEPR process for non-MDA DOD component-directed software, and as appropriate for other DOD component software.*

d. Delegate to appropriate subordinate organizations the authority to release software assets, as appropriate, to the Federal Software Exchange Program (FSEP), reference (d), and for DOD reuse purposes.

e. Specifically address investment strategies, including use of modern software technology and the transition to newer technologies, in the DOD Component planning, programming, and budgeting process.

F.　　UNDERLINE: EFFECTIVE DATE AND IMPLEMENTATION
This Directive is effective immediately.

G.　　REFERENCES:

(a)　　Department of Defense (DOD) Directive 3405.1, "Computer Programming Language Policy," April 2, 1987 (hereby canceled)

(b)　　Assistant Secretary of Defense for Command, Control, Communications and Intelligence (C3I) Memorandum, "Delegation of Authority and Clarifying Guidance on Waivers from the Use of the Ada Programming Language," April 17, 1992 (hereby canceled)

(c)　　DOD Instruction 7930.2, "ADP Software Exchange and Release," December 31, 1979 (hereby canceled)

(d)　　Federal Information Management Regulation (FIRMR) Subpart 201-24.201, Federal Software Exchange Program

(e)　　DOD Directive 8000.1, "Defense Information Management Program," October 27, 1992

(f)　　DOD Directive 5000.1, "Defense Acquisition," March 15, 1996

(g)　　DOD Instruction 5000.2, "Defense Acquisition Management Policies and Procedures," March 15, 1996

(h)　　DOD Directive TS-3600.1, "Information Warfare," December 21, 1992

(i)　　International Organization for Standardization (ISO/IEC 8652:1995), "Ada," February 15, 1995

(j)　　DOD Directive 5200.28, "Security Requirements for Automated Information Systems (AISs)," March 21, 1988

(k)　　DOD Regulation 5200.1-R, "Information Security Program Regulation," December 1987, authorized by DOD Directive 5200.1, "DOD Information Security Program," June 7, 1982

(l)　　DOD 7000.14-R, "DOD Financial Management Regulation," Volume 2, "Budget Presentation and Formulation," May 1994, authorized by DODI 7000.14, "DOD Financial Management Policy and Procedures"

# Appendix B

# Technical Descriptions of Ada and Other Third-Generation Programming Languages

This appendix gives technical and historical descriptions of Ada and the most often cited third-generation programming languages (3GLs)—C and C++—and a new 3GL—Java—the focus of rapidly growing interest in the programming community and a potential candidate for replacing C, C++, or Ada in certain application domains. These descriptions are followed by a comparison of the languages in terms of their capability for ensuring high reliability and for supporting the requirements of long-lived, embedded, real-time, and/or distributed systems.

Any software system can be implemented in essentially any reasonably complete programming language. However, languages vary with respect to how effectively—in terms of cost, schedule, and level of risk—they support the programming of a solution that successfully achieves the required functionality and quality. The descriptions below are intended to clarify such variations among languages. In contrast to the approach taken in the section in Chapter 2 titled "Software Engineering Process and Architecture," architecture, design, and development and maintenance processes here are held fixed; for the purposes of the discussion in this appendix, the only independent variable is programming language choice. As pointed out in Chapter 2, certain programming language choices may enhance the development and maintenance process itself, but that interaction is ignored for the purposes of this comparison.

## ADA 83

Ada 83 was the result of a requirements-driven language design competition, beginning in 1975 with the first "Strawman" requirements document, continuing through a series of requirements documents culminating in the "Steelman" document, and resulting in a preliminary standard in 1980, an American National Standards Institute (ANSI) standard in 1983, and an International Organization for

Standardization (ISO) endorsement of the ANSI standard in 1987. The actual design work was performed by a design team, with review by a panel of experts and the interested public at large.

Major concerns in the design of Ada 83 were reliability, maintainability, human engineering, and efficiency. Human engineering refers to choosing keywords, syntax, and semantics to maximize readability, while trying to minimize "surprise" and error-prone constructs. For example, all control flow constructs have a distinct "end" marker (e.g., "end if," "end loop"), and all program units allow the name of the unit to be repeated (and the compile time to be checked) at the end marker. Parameters may be specified as "in," "out," or "in out" to indicate the direction of information flow upon subprogram call. Formal parameter names may be used at the call point to identify unambiguously the association between formal and actual parameters.

Ada 83 supports strong type checking, extended to provide strong distinctions between otherwise structurally equivalent numeric types, as well as between otherwise structurally equivalent array types and pointer types. Ada 83 is unusual in that it allows the programmer to distinguish two same-sized integer types as representing distinct abstractions, and to specify that an array is meaningfully indexed by one, but not the other, or that a subprogram can meaningfully be passed by one, but not the other. For example, the two integer type declarations

type Month_Number is range 1..12

and

type Hour_Number is range 1..12

introduce two distinct integer types, and the fact that they have identical ranges does not alter the fact that they are distinguishable at compile time when used as array indices, subprogram parameters, and record components. The compiler will detect the use of a value of one type when the other is expected. Furthermore, a change to one, such as switching Hour_Number to be range 0.23, does not have an unintended effect on some other abstraction.

Ada 83 supports data abstraction, modularity, and information hiding through a module construct called a "package" and through "private" types, types whose internal structure is hidden from code outside the defining package. Objects, subprograms, and any other language entity may be declared in the private part or body of a package, thereby hiding it from external access, and allowing revision during maintenance without disturbing external clients of the package.

Program units may be separately compiled while preserving full compile-time consistency checking across units. All program units may have a separate specification and body, allowing the physical configuration control of interfaces to allow productive parallel development of large systems, and enabling interface integrity to be verified before, rather than after, the code is developed.

Packages and subprograms may be defined as "generic" units, which are parameterized by types, objects, and subprograms. Such generic units must be explicitly instantiated with appropriate actual parameters prior to use. Like other units, generic units have a separate specification and body. When a generic unit is compiled, it is checked for legality. Further checks are performed when the unit is instantiated.

Ada 83 defines a complete set of run-time consistency checks to enforce range constraints on numeric types, index constraints on array types, and "discriminate" constraints on other composite types. In addition, all pointers are default initialized to null, and checked for null prior to dereferencing. Ada 83 defines an ability to raise and handle run-time exceptions. The predefined run-time checks all raise such run-time exceptions, allowing the programmer to write fault-tolerant code that catches unanticipated software problems, and performs appropriate recovery or disciplined shutdown actions.

Ada 83 includes a standard multithreading model, with a rendezvous construct to support interthreading communication and synchronization. Explicit delays are supported, as is timed rendezvous.  Finally, Ada 83 includes constructs for explicit user control over representation of types, as well as a "pack" directive to influence the compiler's selection of representation.


## ADA 95

The current Ada standard, Ada 95, was developed between 1990 and 1995.  As with Ada 83, the development was performed by a language design team, and requirements and review were provided through an open forum.  In February 1995, the revised language was approved as an ISO standard, replacing the former edition of the standard.  The overall goal of the Ada 95 design process was to maintain the reliability, maintainability, human engineering, and efficiency of Ada 83, while enhancing the flexibility and extensibility of the language, and the programmer's control over storage management and synchronization.

Ada 95 generalized the type definition mechanisms of Ada 83 to allow a type to be defined as an "extension" of another type, and to treat a type and all its extensions, direct and indirect, as a "derivation class" of types, with "class-wide" operations and dynamically bound implementations of operations. Added to the existing support for abstraction and modularity, type extension and dynamic binding give Ada 95 support for the object-oriented programming paradigm.

Ada 95 also enhanced the multithreading model, by providing "protected objects" that allow the programming of data-oriented synchronization mechanisms, without introducing additional threads.

Ada 95 added support for pointers to subprograms, as well as pointers to declared, as opposed to heap-allocated, objects.  All access types include an "accessibility" level, which is checked by the implementation, generally at compile time, to prevent the creation of dangling references.

The numeric model was enhanced with the addition of modular (unsigned, wraparound) integers with bit-wise logical operators, and decimal fixed-point types, to support exact financial calculations.

The generic facility was enhanced to allow parameterization by packages that are instances of other generics, so that layered generic abstractions may be defined.  In addition, the generic "contract" model was strengthened so that the legality of an instantiation is fully determined by the actual parameters and the generic specification, allowing the body of the generic to be altered during maintenance without endangering the legality of existing instantiations.

Where appropriate, additional run-time checks were defined in Ada 95 to support the enhanced features.  In particular, a conversion from a class-wide type to an extension of its root type involves a run-time check to ensure that the conversion is meaningful, as does a conversion from an "anonymous" access type to a named access type to prevent the creation of a dangling reference (based on the "accessibility" level mentioned above).

In addition to these syntactic and semantic enhancements to the language, a number of additional standard packages, pragmas, and attributes are defined in "annexes" to the standard.  Some of these packages, pragmas, and attributes must be supported by all implementations, such as packages for string manipulation and random number generation and pragmas for interfacing to other languages.  Others are specifically designed to support particular application domains, such as real-time, distributed systems, and safety/security-critical systems.


## C

The C language was designed at Bell Laboratories in the early 1970s, as a successor to the language BCPL, for the purpose of writing an operating system (Unix) and associated utilities for minicomputers.  During the late 1970s, C and Unix were used widely in universities, and during the 1980s C emerged as the language of choice for systems programming on minicomputers, workstations, and personal computers. The ANSI standard for C was approved in 1989, and the ISO standard based on ANSI C was approved in 1990.

C has a sparse syntax, with braces used for begin and end markers in all control flow, program unit, and type declaration constructs. Single-character operators are provided for assignment, indirection, address-of, bit-wise and, or, "xor," and not, and the usual arithmetic operations. Operators are also provided for pre- and post-increment and decrement, operate-and-assign, and left- and right-shift.

Numeric data types are selected by names, such as "short int" or "long float." There is no capability to select a numeric type by required range or precision, and there is no notion of implementation-enforced range constraint. Enumeration data types are supported, but are implicitly convertible to and from integer types in any context.

Historically, interface definitions have not been necessary for C functions, with the default being that a function returns an "int" and takes any number of parameters. ANSI C introduced the notion of a function "prototype" to specify the function interface, and some implementations can be directed to require the presence of a prototype for all functions.

All arrays are indexable by any integer or enumeration type; all arrays have a low bound of zero, and a high bound of one less than the specified size. No bounds information is carried with array parameters, and no bounds checking is defined by the language standard, although some tools exist that will check for out-of-bounds references. Arrays are treated by the language as essentially constant-valued pointers, and array indexing is defined in terms of an indirection applied to the addition of a pointer and an integer index.

Strings in C are represented by a pointer to their first character, with a null character used by convention to signify the end of the string. There is no language-defined checking for running off the end of a string.

Record-like "structs" are supported, but there is no language-defined data abstraction mechanism. Opaque, incomplete pointer types can be used to provide some degree of data abstraction. A "union" construct allows the creation of an undiscriminated union of types. There is no language-defined check for accessing the "wrong" member of a union.

The "cast" construct may be used to explicitly convert between numeric types (although implicit conversion is performed as part of a function call, and implicit widening is performed during arithmetic). The cast construct may also be used to convert between pointer types, or between an integer and a pointer type. There is no language-defined check associated with a cast.

No default initialization is defined by the standard for local variables; pointers, in particular, are not default initialized. No null-checking is defined for pointer indirection.

There is no language-defined construct for raising and handling exceptions, although there are standard functions for sending and handling "signals," which can be used to emulate exceptions in certain circumstances.

C provides some control over representation by the use of bit field indicators on "struct" components. However, it does not define the ordering of bit fields within a word. Some implementations provide "pack" pragmas or other means of providing more representation control.

There is no language-defined "module" construct other than a source file; objects and functions declared "static" are local to the source file. Objects and functions not declared "static," when defined at the top level, are externally visible from any other file that includes an "extern" declaration for the entity. By convention, the "extern" declarations for a source file, and associated type definitions, are usually grouped into a header file (".h" file), which can be textually included ("#include") in any source file requiring access to the type, object, or function.

C includes a standard preprocessor that supports textual include, conditional compilation, and parameterized textual macros.

The ANSI C standard includes a full set of library functions to support string manipulation (where a string is a null-terminated array of characters), random number generation, and input/output, among others.


# C++

The C++ language was first released in 1983 as an enhancement to C, with the major enhancement being the addition of a "class" construct inspired by the same-named feature of the language Simula-67.  The language was initially defined by the implementation available from AT&T ("cfront") that translated C++ to C.  Cfront, and hence C++, went through several major updates that added features such as multiple inheritance, generic templates, and exception handling.  In the early 1990s, ANSI and ISO committees were formed to produce a standard for the language.  A few additional features, such as run-time type identification and "namespaces," have been added during the standardization process.  Approval of the ISO C++ standard is expected within the next year.

C++ includes all the features of C, although some features are revised to be more strongly typed.  For example, enumeration types in C++ are implicitly convertible to integer types, but not implicitly convertible back.  Also, function prototypes are required for all C++ functions.  C++ adds to C support for data abstraction, type inheritance, and dynamic binding (virtual functions).  Two kinds of multiple inheritance are supported: the default inheritance replicates the fields if the same base class is inherited through multiple paths, and the virtual inheritance shares fields if the same base class is inherited through multiple paths.

C++ also supports a generic template facility.  No checking is defined for templates prior to instantiation; there is no template "contract" model.  Instantiation is implicit by referring to an instance via "template_name<parameters>."  Template functions are also supported; instantiation of a template function is automatic at a call, with the template parameters determined implicitly by the types of the call parameters.

C++ supports "throw"ing and "catch"ing exceptions.  Exceptions can be represented by objects of any type; the "catch" is based on a type matching.  The standard C++ library defines certain exception types, instances of which are thrown when an allocator fails to allocate storage, or when other errors occur.

The array indexing and cast constructs inherited from C remain unchecked in C++.  There are standard templates for defining checked arrays and checked casts.  Local pointers in C++ are not default initialized, and there is no language-defined check for dereferencing a null pointer.  "Smart pointer" abstractions can be developed to check for null pointers, or to implement persistence or similar capabilities.

As in C, all numeric types are implicitly convertible on assignment and parameter passing, and implicitly "widened" in calculations.

C++ supports information hiding through the notion of protected and private data and function members.  Private members are visible only inside a class (and to its "friends").  Protected members are visible inside all descendants of a class.  C++ supports a multilevel namespace though a "namespace" construct, which provides no information hiding (there is no "private" part of a namespace) and is simply a hierarchical naming mechanism.


# JAVA

Java was developed over the past 5 years at Sun Microsystems.  It was originally called "Oak" and was intended for use in small appliances, set-top boxes, and other embedded applications.  In April

1995, a World Wide Web browser written in Java, called HotJava, was announced by Sun. HotJava had the ability to download small programs written in Java over the Web and execute them in the context of a Hypertext Markup Language (HTML) page being displayed by the Web browser. Since then, Sun's Java technology has been licensed by essentially all other Web browser developers, including Netscape and Microsoft, and has achieved widespread attention for its potential to provide many of the capabilities of client/server systems without many of the attendant complexities.

Java is syntactically based on C++ but semantically is closer to Modula-3 or Ada 95. It provides modularity through a combination of a "package" concept, which is a namespace with some information hiding associated with it, and the "class" construct, which is modeled closely on the C++ (and Simula-67) class construct. To support information hiding, methods (called "member functions" in C++) and data components may be marked as public, protected, or private, much as in C++, but with the added notion that, by default, methods and data are visible only to classes within the same package. Unlike C++, there is no textual "include" in Java; instead, individual classes, or a whole package of classes, are explicitly imported using an "import" statement at the top of the source file defining a class.

All code and objects in Java must be inside some class. The methods of a class are by default "virtual" in Java; calls to such methods are "dynamically" bound. Methods may be explicitly specified as "static"; calls to such methods are "statically" bound. The data components of a class are by default "per-instance," as in C++. Data components may be marked "static," which means that they are "statically" allocated and shared across the class, rather than one per instance.

Java fully supports single inheritance between classes. By default a class inherits from the single "root" type called "java.lang.Object." Alternatively, it may explicitly specify one parent class from which it inherits non-static methods and data components. Java provides a limited kind of multiple inheritance through the concept of an "interface" type: a list of methods that any "implementor" of the interface must provide. A class may specify any number of interface types that it claims to "implement." The compiler verifies that the methods required by each identified interface are present in the class. There is no separate specification for a class (other than that provided by interface types it implements). There is no separate "prototype" for a method of a class. A tool may be used to extract the documentation and specification for class.

Java has no direct support for enumeration types. Named integer constants may be used, but the compiler provides implicit widening between integer types on assignment and parameter passing and allows any integer type to index any array. Arrays in Java are indexed from zero, as in C and C++, but unlike C or C++, their semantics are not defined in terms of pointer arithmetic. In fact, Java does not support pointer arithmetic. Arrays are first class types, and carry a length at run time against which all indexing is checked.

Pointers ("references") in Java are default initialized to null, and all pointer dereferences are checked for null. Conversions between references are checked at run time for meaningfulness.

Java has exceptions, much as in C++, except that it enforces compatibility and completeness of "throw" signatures at compile time (C++ enforces "throw" signatures at run time). Failures of run-time checks, such as an array-bounds check, or a null-pointer check, result in a "throw" of a predefined exception. Run-time error exceptions do not need to be mentioned in a "throw" signature; other exceptions, including user-defined exceptions, do need to be mentioned in the "throw" signature of a method if it is going to throw or propagate the exception.

Java has no generic templates; the root type java.lang.Object can be used in some contexts to define (heterogeneous) "generic" data structures. Proposals exist to add a parametric polymorphism facility to Java, which could provide some of the added compile-time type checking associated with "homogeneous" data structures provided by the generic template features of Ada and C++.

Java has largely the same control flow constructs as C++. As in C and C++, switch statements rely on a programmer-inserted "break" to terminate a case. Java defines a special type "boolean" and requires a value of boolean type in the expression of an "if," "while," or "for" test. There is no implicit conversion to "boolean"; the relational operators return boolean, as do the logical operators.  The operator "=" is for assignment; "==" is for equality.  A "break" or "continue" statement may have an identifier to identify the particular construct being exited or continued, providing some additional flexibility and maintainability relative to C and C++.

All class and array instances are allocated dynamically on a garbage collected heap.  That is, there are no "stack-resident" arrays or class instances.  All implementations of Java provide a garbage collector.  There are no class instances "nested" inside other class instances, only references to dynamically allocated class instances.  The same goes for arrays.

There is no user control over representation of data objects.  There is no user control over storage management, other than a system method to force a garbage collection.

Java has a large standard library of classes and includes support for multithreading through a combination of a standard thread class and the notion of "synchronized" methods.


## DOMAIN-SPECIFIC COMPARISON


In comparing the features of Ada, C, C++, and Java, various principles underlying each language can be identified.

With C, the underlying goal is to provide reasonable portability (certainly when compared with assembly language) while giving the programmer full control of the machine.  There is little attempt to provide strong consistency checking at compile time, and no notion whatsoever of run-time checking built into the language (other than via use of the standard "assert" macro).

C++ provides more tools for defining abstractions, and increases the strength of the type checking on enumeration types.  However, the default run-time behavior in C++ is still inherited from C, which means no run-time initialization or checking of pointers, no checking of array indexing, and no notion of range checking.  The default conversion syntax, the simple "cast" inherited from C, does no checking.  The basic primitives of C++ remain unsafe, although there are additional mechanisms available for creating safe abstractions.

Java takes the route of strongly enforcing run-time consistency, with all the necessary checks to ensure that a program does not corrupt data outside its prescribed space, including pointer initialization and null checking, array-bounds checking, and conversion checks.  However, at compile time, Java has essentially gone one step backward from C++ by dropping support for enumeration types, thereby eliminating an important source of compile-time consistency checks.  Java very successfully creates a language that prevents code from corrupting data outside its purview, but it fails to provide tools for supporting thorough compile-time enforcement of interface consistency.

A second area of concern with regard to use of Java for critical systems development is that it is inextricably tied to a dynamic storage allocation model.  Garbage collection is certainly less error-prone than is manual storage reclamation, but any use of dynamic storage allocation opens up the possibility of eventual storage exhaustion, as does dynamic stack extension.  For an embedded or critical system, it is standard practice to require that all storage be allocated statically (at link time), including the stacks for all threads of control; recursion is also disallowed.

In comparison to the above languages, Ada 83 and Ada 95 attempt to provide more features to make compile-time consistency checking useful for finding mistakes, backed up by run-time consistency checks for cases in which only a dynamic check is meaningful.  As mentioned above, Ada is one of the few languages that allows the programmer to create strong distinctions between structurally equivalent numeric, array, and pointer types.  These distinctions allow an Ada interface to capture more of the semantics, and allow the Ada compiler to catch more mistakes in the use of an interface.  The last decade has seen an explosion in the number of application programming interfaces (APIs) used to build systems.  Inappropriate uses of an API are among the most common mistakes in such systems.  By creating stronger distinctions between numeric, enumeration, array, and pointer types, an Ada version of an API can reduce the likelihood of inappropriate use, and identify more such errors at compile time.

At run time, Ada has pointer default initialization, pointer null checking, array bounds checking, with user control over both the low and high bound, and conversion checking. In addition, Ada provides range checking, variant record checking, and, in Ada 95, both compile-time and run-time checks designed to eliminate "dangling" references associated with pointers to deallocated stack variables. This set of "dangling reference" checks ("accessibility checks") allows an embedded or critical program to avoid completely the use of dynamic storage allocation, while still providing the convenience of using pointers.

Both Ada and Java have support for multithreaded applications as a standard, portable part of the language, whereas C and C++ support multithreading generally through operating-system-dependent libraries. The Ada multithreading support includes various real-time-oriented features, such as timed entry calls and selective accepts with delay alternatives, whereas Java has only a basic timed "sleep" operation. To the basic Ada 83 multithreading support, Ada 95 adds protected objects, which are designed to support real-time systems by reducing overhead, minimizing "priority inversion," and generally improving predictability of thread synchronization. Java's synchronized methods, with wait/notify operations, provide similar capability, although with less encapsulation of the fields requiring synchronized access, a more race-prone "notification"-oriented synchronization model, and no particular concern for priority inversion.

Although Ada is a general-purpose 3GL, it was designed with extra attention to the concerns of real-time, embedded, and critical systems developers, namely very thorough consistency checking, mechanisms to support a very "static" storage allocation model, and multithreading support with time- and priority-cognizant constructs. As such, at a technical level, it is a better fit to the needs of DOD critical and embedded systems development than are the other languages in widespread commercial use. These reliability-oriented features of the Ada language make development and maintenance more cost-effective, when cost to achieve the required level of quality and correct functionality is included. Of course, there are other non-technical issues involved in language choice (as discussed in Chapter 1), and other non-language issues involved in managing successful software development (discussed in Chapter 2).

# Appendix C

# Glossary

**Application Programming Interface (API)**. A set of procedure and function specifications providing access to the capabilities of a reusable software component, such as a "windowing" or network communication operating subsystem.

**Architecture**. The structure of a system's components and connectors, their interrelationships, and the principles and guidelines governing their design and evolution over time.

**Automated Information System (AIS)**. A combination of computer hardware and computer software, data, and/or telecommunications that performs functions such as collecting, processing, transmitting, and displaying information. The function of such systems is primarily administrative. Excluded are computer resources, both hardware and software, that are physically part of, dedicated to, or essential in real time to the mission performance of weapon systems; used for weapon system specialized training, simulation, diagnostic testing and maintenance, or calibration; or used for research and development of weapon systems.

**Commercial Item**. An item regularly used in the course of normal business operations for other than government purposes that has been or will be sold or licensed to the general public, and that requires no unique government-directed modifications or maintenance over its life cycle to meet the needs of the procuring agency.

**Commercial Off-the-Shelf (COTS) Software**. COTS software products are commercial items that have been sold, leased, or licensed in a quantity of at least 10 copies in the commercial marketplace, at an advertised price. COTS software products include a description or definition of the functions the software performs, documented to good commercial standards, and a definition of the resources needed to run the software.

**Domain**. A distinct functional area that can be supported by a class of systems and assets with similar requirements and capabilities.

**Fourth-Generation Programming Language (4GL)**. A 4GL differs from a third-generation programming language (3GL) in that it removes the need for a programmer to explicitly make many of the design decisions about data structures and algorithms. 4GLs allow programmers to express instructions in terminology and at a level of abstraction that are natural for communication between humans who are familiar with the application domain. The programmer interaction may even be in a graphical, or menu-based, form. 4GLs typically incorporate domain-specific knowledge and notation, and so are not "general purpose" in the sense that 3GLs are, although they may include a 3GL component that permits general-purpose programming. 4GLs are associated with "frameworks," "templates," "automatic program generators," "middleware," and "graphical user interface builders." Some examples of 4GLs are Visual Basic, PowerBuilder, Delphi, and SQL.

**Function Point**. A measure of software functionality that is independent of differences in lines of code required to implement a given function in different programming languages.

**Glue Code**. The relatively small parts of computer programs, or operating system "shell" scripts, that are written to integrate non-developmental items into a larger system, without modification to the components themselves.

**Government-directed Software Maintenance**. Maintenance required for software changed in response to government specification or direction. Government-directed maintenance may be done by the government or by a commercial organization paid by the government. In contrast, with vendor-directed maintenance, a vendor chooses which changes are made to the software, and when they will be made.

**Non-Developmental Item (NDI)**. Any software asset that is available in the commercial marketplace; or any software asset that is available to the public for free use; or any previously developed software asset that is in use by a department or agency of the United States, a state or local government, or a foreign government with which the United States has a mutual defense cooperation agreement.

**Product Line**. A set of similar products or a family of systems that share common architectures and satisfy the mission requirements of one or more domains.

**Re-engineering**. The process of examining and altering an existing system to reconstitute it in a new form. Re-engineering may include reverse engineering, restructuring, re-documenting, forward engineering, re-targeting, or translation.

**Reuse**. The process of developing or updating a software-intensive system using existing software assets.

**Software Asset**. Any software-related product of the software life-cycle.

**Software Engineering Environment**. The set of tools (including supporting hardware, software, and "firmware") used in the production and maintenance of software throughout its life-cycle. Typical elements include computer equipment, compilers, operating systems, "debuggers," simulators, emulators, computer-aided software engineering tools, and database management systems.

**Software Maintenance**.  Maintenance of software includes, but is not limited to, activities generally referred to as enhancement, evolution, post-deployment software support, or error correction.

**Software Metrics**.  Quantitative values used to make an assessment of software condition, products, or processes.  Representative metrics are effort, schedule, cost, quality, size, and rework.

**Standards-based Technical Architecture**.  An architecture that defines the standards, services, topology, data definitions, and common framework that enable systems developed to the architecture to interoperate.

**Third-Generation Programming Language (3GL)**.  Third-generation programming languages generally differ from second-generation languages in being machine-independent, providing built-in control structures, and supporting user definition of abstractions, including subprograms and data types. They differ from fourth-generation languages in continuing to require the programmer to deal explicitly with the design of data structures and algorithms.  Some examples of 3GLs are Ada, C, C++, Cobol, Fortran, Java, Jovial, Pascal, and SmallTalk.

**Warfighting Software**.  There are two primary criteria for determining whether a subsystem belongs in the "warfighting" category:

1.  Relatively little commercial software and expertise is available for implementing the desired functions.  For example, even though intelligence analysis is involved in warfighting, many of its functions (database update, query and visualization, report generation) can be readily satisfied via non-warfighting commercial software.

2.  The application requires software quality attribute levels higher than those supportable by commercial software.  For many warfighting functions, these involve real-time performance, reliability, and survivability, particularly in high-stress, crisis-mode situations in which DOD information processing functions may be under attack.

The application domains for warfighting software include, but are not necessarily limited to, the following areas:

• *Weapon control*, which includes software involved in weapon sensor processing; guidance, navigation, and control; combat-oriented weapon-delivery platform control; and software for special weapon delivery platform operator devices such as heads-up displays.  Weapon control does not include administrative functions and "hotel services" for large weapon delivery platforms such as aircraft carriers, or support subsystems performing mainstream data management, networking, and graphical user interface functions.

• *Electronic warfare*, which includes software involved in rapid-response electronic detection, identification, discrimination, tracking, platform-based communication, and associated countermeasure/counter-countermeasure applications, but does not include support subsystems performing mainstream data management, networking, and graphical user interface functions.

• *Wideband real-time surveillance*, which includes software involved in hard or soft real-time image, infrared, radar, or other sensor processing, but does not include off-line query and analysis of surveillance archives or support subsystems performing mainstream data management, networking, and graphical user interface functions.

• *Battle management and battlefield communication*, which includes hard or soft real-time weapons allocation, targeting, control, coordination, damage assessment, and associated battlefield communications requiring such special capabilities as spread spectrum, anti-jamming, and frequency-hopping, but does not include off-line monitoring, update, query, and analysis of battle asset status, or off-battlefield communications. Thus, the range of "warfighting command, control, and communications ($C^3I$) applications" is narrower than previous categorizations such as "$C^3I$" or "mission critical."

The scope of warfighting applications also does not include associated support software for test, simulation, training, off-line analysis, maintenance, and diagnostics.

# Appendix D

# Detailed Comparisons of Ada and Other Third-Generation Programming Languages

This appendix presents details of studies, summarized in Chapter 2 in the sections titled "Analyses of Language Features" and "Comparisons of Empirical Data," that have analyzed the technical features of programming languages and have used empirical project data to compare languages' relative strength with respect to desired characteristics. Given the unscientific nature of most of the data, no strong conclusions are warranted. But in a general sense, most experience with Ada supports the conclusion that it benefits DOD warfighting applications; few data were found to refute such a conclusion, although the absence of such data may reflect a lack of any organized effort to promote the use of languages other than Ada in warfighting systems.

## EVALUATIONS OF LANGUAGE FEATURES

A 1985 Federal Aviation Administration (FAA) study compared Ada with four other languages by conducting expert evaluations based on 48 technical language features arranged in six categories, and combining the evaluation results with the results of performance benchmark test results (IBM, 1985). A follow-on study was conducted by the Software Engineering Institute (SEI) in 1991 using the same expert evaluation methodology as the original study to compare Ada 83 with C++, but without the benchmark tests (Weiderman, 1991). The studies, the results of which are presented in Table D.1, compared the languages against the maximum possible scores for each of the six categories, which can be summarized as follows:

- *Capability*—facets of the implementation language relevant to programming or software engineering;
- *Efficiency*—factors relevant to optimization of generated code and run-time utilization of resources;

Table D.1  Language Functionality Attributes

| Language Attributes | Maximum Score | FAA/IBM Study | | | | | SEI Study | |
|---|---|---|---|---|---|---|---|---|
| | | Ada 83 | C | Pascal | Jovial | Fortran | Ada 83 | C++ |
| Capability | 16.7 | 16.1 | 9.6 | 10.4 | 7.6 | 3.9 | 15.3 | 11.3 |
| Efficiency | 16.4 | 8.0 | 11.8 | 10.8 | 11.0 | 11.1 | 10.7 | 10.9 |
| Availability/ Reliability | 22.6 | 21.5 | 11.6 | 14.5 | 15.6 | 10.3 | 19.1 | 12.6 |
| Maintainability/ Extensibility | 17.4 | 14.0 | 10.2 | 12.2 | 6.8 | 8.3 | 13.6 | 11.4 |
| Life-cycle cost | 11.3 | 8.2 | 7.4 | 7.8 | 4.9 | 5.2 | 8.4 | 8.0 |
| Risk | 15.6 | 8.8 | 8.9 | 7.6 | 9.6 | 8.2 | 11.7 | 9.8 |
| Total | 100.0 | 76.6 | 59.5 | 63.3 | 55.5 | 47.0 | 78.8 | 64.0 |

SOURCES:  Federal Aviation Administration (FAA)/IBM data from IBM (1985); Software Engineering Institute (SEI) data from Weiderman (1991).

- *Availability/Reliability*—factors influencing day-to-day safety of operational systems;
- *Maintainability/Extensibility*—factors influencing long-term viability of operational systems;
- *Life-cycle cost*—elements of cost associated with or affected by the implementation language; and
- *Risk*—areas of uncertainty or concern associated with or affected by the implementation language.

Ada scored the highest overall in both studies.  It scored the highest in each individual category, except for efficiency, where it was ranked last, and risk, where it was ranked first in the SEI study and third in the FAA study.

## PROJECT EVALUATION DATA

In comparisons of Ada with other programming languages in actual development projects, empirical data on characteristics such as expressibility, maintainability, defect rates, reusability, and cost factors are of particular interest.  Since Ada has been used in relatively large programs, a larger proportion of data has been collected for Ada than for contemporaneous commercial programming languages like C and C++.

Much of the data are subjective in nature, and care must be taken in drawing conclusions from such studies.  For instance, the data in Table D.2 are confounded by differences in the expressive power of a source line of code in different programming languages.  One way of normalizing uses tables of source lines of code per function point (Jones, 1995).  However, as shown in Table D.2, these ratios have wide variability.  Lubashevsky (1996) reports variations in source lines of code per function point exceeding factors of 2 for C and 6 for C++.  Finally, there are differences in expressiveness for the same language across different application domains.

Table D.2  Source Lines of Code per Function Point

| Language | Low | Mean | High |
|----------|-----|------|------|
| Ada 83 | 60 | 71 | 80 |
| C | 60 | 128 | 170 |
| C++ | 30 | 53 | 125 |

SOURCE:  Data from Jones (1995).

## Open-Source Data

The Verdix Ada Development System (VADS) data (Zeigler, 1995) covers a period of approximately 10 years of development and enhancement of VADS compilers, debuggers, builder tools, and run-time support systems using both C and Ada.  Table D.3 summarizes the comparative C and Ada experience data over this period.

Zeigler analyzed potential confounding effects of relative C and Ada software complexity, personnel capability, and learning curve effects and found that these factors did not cause any significant bias.  The study thus provides a strong case that Ada was significantly better than C, in terms of both life-cycle cost and freedom from defects, for this large (over 1 million lines of both Ada and C) project in the compiler and tools area.  However, the study can only be suggestive rather than definitive about the applicability of this result to other domains and other teams (the Verdix teams were both composed of high-capability, low-turnover personnel).

Since 1976, NASA's Software Engineering Laboratory (SEL) has been carefully collecting software data on non-embedded, non-real-time satellite flight dynamics software.  Figure D.1 (McGarry et al., 1994) summarizes SEL's history of development error rates over this period.  The data show a significant general decrease in error rates over the period, due to such techniques as code reading, systematic testing, and Cleanroom techniques.  The data from 1988 to 1994 also show significantly lower error rates for Ada than for Fortran, measured per 1,000 lines of delivered code, or KDLOC (an average of 2.1 defects/KDLOC for Ada vs. 4.8 defects/KDLOC for Fortran).  SEL's analyses for Ada initially concluded that owing to reuse, Ada projects experienced significant cost and schedule reductions compared to Fortran projects.  Subsequently, however, the Ada object-oriented reuse approach was applied to Fortran projects, resulting in comparable gains.

## Proprietary Data

This section summarizes proprietary data from Reifer (1996) and Jones (1994) that compare Ada to other programming languages with respect to cost, delivered defects, reliability, and productivity.  Because the source data and detailed analysis are unavailable, these results are difficult to assess.

*Cost*

Data on cost, reliability, and maintainability have been compiled in a proprietary database of outcomes of 190 software projects 3 years old or less (Reifer, 1996).  These projects were written in Ada 83, Ada 95, C, C++, and other third-generation programming languages (3GLs, such as Cobol or Fortran) and were compared to a norm for each domain, drawn from a database of over 1,500 projects 7 years old or less.

Table D.3  Verdix Ada Development System Development Data

| Characteristic | C | Ada | Script Files | Other Files | Totals |
|---|---|---|---|---|---|
| Total lines | 1,925,523 | 1,883,751 | 117,964 | 604,078 | 4,531,316 |
| SLOC[a] | 1,508,695 | 1,272,771 | 117,964 | 604,078 | 3,503,508 |
| Files | 6,057 | 9,385 | 2,815 | 3,653 | 21,910 |
| Updates | 47,775 | 34,516 | 12,963 | 12,189 | 107,443 |
| New features | 26,483 | 23,031 | 5,594 | 6,145 | 61,253 |
| Fixes/feature | .52 | .25 | .82 | .17 | .41 |
| Fixes/KSLOC[b] | 9.21 | 4.59 | 39.02 | 1.75 | 7.25 |
| Cost/SLOC | $10.52 | $6.62 | $15.38 | $3.72 | $8.10 |
| Defects/KSLOC | .676 | .096 | n/a | n/a | .355 |

[a] SLOC:  Source lines of code.
[b] KSLOC:  1,000 SLOC.
SOURCE:  Zeigler (1995).

Reifer's project cost data, shown in Table D.4, indicate that, in general, Ada, C, and C++ programs were less expensive per delivered source line of code (SLOC) than other 3GL programs or the domain norm for military applications, although C and C++ programs were less expensive than Ada programs in all military domains except airborne and spaceborne weapons systems.  In commercial domains, the development cost of C and C++ code was again less than or equivalent to the cost of Ada, and Ada was roughly equivalent in cost to other 3GLs and the domain norms.  However, the wide variations in expressive power of a line of code in Ada, C, and C++ (see Table D.2) tend to overshadow the cost/SLOC differences shown in Table D.4.

Another study used telecommunications systems to evaluate object-oriented programming languages (Jones, 1994).  The study used as a baseline a private branch exchange switch software application of 1,500 function points in size, compared across several telecommunications companies.  Based on these data, Table D.5 shows the cost of developing the project in different languages, in dollars per function point.  The results indicate that, in this domain, coding in C++ was less expensive than coding in Ada, while using C was more expensive.

*Delivered Defects*

Reifer's (1996) data, shown in Table D.6, indicate that in the 190 software projects represented, Ada code had fewer delivered defects than C and C++ code, which, in turn, had fewer defects than other 3GLs.  In command and control and telecommunications applications, the difference between Ada and C/C++ was smaller than the difference between C/C++ and other 3GLs and the norm.  In commercial products, information systems, and weapons systems, Ada code clearly had fewer defects than code in C, C++, and other 3GLs.

Based on Jones's (1994) data, Table D.7 indicates that, for telecommunications applications, the reliability in delivered defects per function point was roughly equivalent for C++ and Ada, and higher for C code.

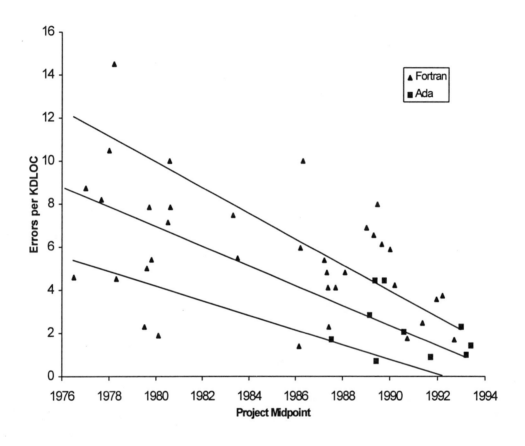

FIGURE D.1 NASA-SEL development error rates. KDLOC: 1,000 delivered lines of code. (Reprinted by permission from McGarry et al. (1994). Copyright 1994 by Carnegie Mellon University.)

Table D.4  Cost Data by Language (dollars per delivered source line of code)

| Application Domain | Ada 83 | Ada 95 | C | C++ | 3GL | Norm |
|---|---|---|---|---|---|---|
| Command & Control | | | | | | |
|   Commercial | 50 | n/a | 40 | 35 | 50 | 45 |
|   Military | 75 | n/a | 75 | 70 | 100 | 80 |
| Commercial Products | 35 | 30 | 25 | 30 | 40 | 40 |
| Information Systems | | | | | | |
|   Commercial | n/a | n/a | 25 | 25 | 30 | 30 |
|   Military | 30 | 35 | 25 | 25 | 40 | 35 |
| Telecommunications | | | | | | |
|   Commercial | 55 | n/a | 40 | 45 | 50 | 50 |
|   Military | 60 | n/a | 50 | 50 | 90 | 75 |
| Weapons Systems | | | | | | |
|   Airborne & Spaceborne | 150 | n/a | 175 | n/a | 250 | 200 |
|   Ground-based | 80 | n/a | 65 | 50 | 100 | 75 |

SOURCE: Reprinted from Reifer (1996).

Table D.5  Telecommunications Project Costs by Language (dollars per function point)

| | Assembly | C | CHILL | Pascal | Ada 83 | Ada 9X[a] | C++ | Smalltalk |
|---|---|---|---|---|---|---|---|---|
| Cost[b] | 5,547 | 2,966 | 2,260 | 1,993 | 1,760 | 1,533 | 1,180 | 1,007 |

[a] Ada 9X figure based on simulation.
[b] At a monthly salary rate of $10,000.
SOURCE:  Data from Jones (1994).

The fairly consistent factor-of-10,000 relationship between the data on cost per function point shown in Table D.5 and the data on defects per function point shown in Table D.7 suggests that Jones's cost and defect data are highly correlated to some third factor.  This third factor appears to be the mean SLOC/function point ratio from Jones (1995), shown in Table D.2.  Comparing these ratios across C, Ada, and C++ indicates that they are approximately, although not exactly, equal, as shown in Table D.8.

*Reliability*

Data from Reifer (1996) in Table D.9 indicate that the incidence of failure for Ada programs was lower than that with C, C++, and other 3GLs.  The outcome of this reliability comparison is similar to the defect comparison; Ada was superior to C and C++, which were better than other 3GLs.

Table D.6  Delivered Defect Data by Language (post-delivery errors per KSLOC[a])

| Application Domain | Ada 83 | Ada 95 | C | C++ | 3GL | Norm |
|---|---|---|---|---|---|---|
| Command & Control | | | | | | |
|   Commercial | 1.5 | n/a | 1.9 | 1.5 | 2.7 | 2.5 |
|   Military | 0.7 | n/a | 1.0 | 1.2 | 2.3 | 2.0 |
| Commercial Products | 2.8 | 3.5 | 5.0 | 3.0 | 4.5 | 4.0 |
| Information Systems | | | | | | |
|   Commercial | 4.0 | n/a | 7.0 | 5.1 | 7.0 | 7.0 |
|   Military | 3.0 | n/a | 6.0 | 4.0 | 6.0 | 6.0 |
| Telecommunications | | | | | | |
|   Commercial | 1.6 | n/a | 2.0 | 1.7 | 3.3 | 3.1 |
|   Military | 1.0 | n/a | 1.5 | 1.2 | 2.7 | 2.5 |
| Weapons Systems | | | | | | |
|   Airborne & Spaceborne | 0.3 | n/a | 0.8 | 0.6 | 1.0 | 1.0 |
|   Ground-based | 0.5 | n/a | 0.8 | 0.7 | 1.0 | 1.0 |

[a] KSLOC: 1,000 source lines of code.
SOURCE: Reprinted from Reifer (1996).


Table D.7  Defects for Telecommunications Applications (delivered defects per function point)

| | Assembly | C | CHILL | Pascal | Ada 83 | Ada 9X[a] | C++ | Smalltalk |
|---|---|---|---|---|---|---|---|---|
| Defects | 0.52 | 0.29 | 0.23 | 0.20 | 0.17 | 0.15 | 0.14 | 0.13 |

[a] Ada 9X figure based on simulation.
SOURCE:  Data from Jones (1994).


Table D.8  Correlation in Function Point (FP) Measures

| | SLOC[a]/FP (Jones, 1995) | Dollars/FP (Jones, 1994) | Defects/FP (Jones, 1994) |
|---|---|---|---|
| C/Ada | 1.80 | 1.69 | 1.71 |
| C++/Ada | 0.75 | 0.67 | 0.82 |

[a] SLOC:  Source lines of code.

Table D.9  Reliability Data by Language (weeks until next major repair incidence)

| Application Domain | Ada 83 | Ada 95 | C | C++ | 3GL | Norm |
|---|---|---|---|---|---|---|
| Command & Control | | | | | | |
| Commercial | 3.0 | n/a | 2.5 | n/a | 2.0 | 2.5 |
| Military | 4.0 | n/a | 3.0 | n/a | 2.0 | 2.5 |
| Commercial Products | 1.0 | n/a | 0.4 | 1.0 | 0.4 | 0.5 |
| Information Systems | | | | | | |
| Commercial | 1.0 | n/a | 0.5 | 0.6 | 0.4 | 0.5 |
| Military | 0.8 | n/a | 0.5 | n/a | 0.4 | 0.5 |
| Telecommunications | | | | | | |
| Commercial | 3.0 | n/a | 1.0 | 2.0 | 1.5 | 1.8 |
| Military | 4.0 | n/a | 2.0 | 3.0 | 2.0 | 2.0 |
| Weapons Systems | | | | | | |
| Airborne & Spaceborne | 8.0 | n/a | 3.0 | n/a | 2.5 | 2.5 |
| Ground-based | 6.0 | n/a | 3.0 | n/a | 2.0 | 2.5 |

SOURCE: Reprinted from Reifer (1996).

*Productivity*

As described in Chapter 2 in the section titled "Software Engineering Process and Architecture," one of the key relationships in software development productivity is the relationship of development effort to program size—that is, the value of the "process exponent." A MITRE study (Giallombardo, 1992) found substantial improvements in productivity for large embedded military software developments provided by Ada compared with other 3GLs. The study found that effort, as measured by staff-months required per 1,000 equivalent delivered source instructions (KEDSI; which counts terminal semicolons rather than source lines of code), increased linearly for the large Ada projects (process exponent close to 1). For large non-Ada projects, effort increased more than linearly; the increase in effort was also greater than that predicted by the COCOMO model, which has a process exponent of 1.2 for embedded or very large systems (Boehm and Royce, 1988). The differences were clearest for system sizes greater than 100 KEDSI (which the MITRE study used as a definition of "large").

In a study of telecommunications applications, Jones (1994) compared the productivity of several 3GLs, using function points per staff-month as the metric. The results, presented in Table D.10, indicate that, in this domain, C++ developments had a higher productivity rate than those using Ada, while developments using C were less productive than those using Ada.

Table D.10  Productivity for Telecommunications Projects by Language (function points per staff-month)

| | Assembly | C | CHILL | Pascal | Ada 83 | Ada 9X[a] | C++ | Smalltalk |
|---|---|---|---|---|---|---|---|---|
| Productivity | 1.80 | 3.37 | 4.42 | 5.01 | 5.68 | 6.52 | 8.47 | 9.99 |

[a] Ada 9X figure based on simulation.
SOURCE: Data from Jones (1994).

## CONCLUSION

In summary, based on the results of currently available empirical data and feature analysis comparisons, a conclusion that Ada is superior, with respect to availability/reliability and a lower incidence of defects, appears warranted.  The evidence is not strong enough to assert Ada's superiority in the cost area.  However, there is some positive evidence in the cost area, and when it is combined with the anecdotal  conclusions favoring Ada (described in "Anecdotal Experience from Projects" in Chapter 2) and the lack of solid evidence indicating less expensive custom software development in other languages, a case can be made that using Ada provides cost savings in building custom software, particularly for real-time, high-assurance warfighting applications.

# Appendix E

# Briefings and Position Papers Received by the Committee

**BRIEFINGS**

**Washington, D.C., April 10-12, 1996**

1. "Charge to the Committee," Emmett Paige, Jr., Assistant Secretary of Defense (Command, Control, Communications, and Intelligence), and Cynthia Rand, Principal Director for Information Management, Office of the Assistant Secretary of Defense (Command, Control, Communications, and Intelligence).

2. "Ada Policy and Reality," Charles Engle, Director, Ada Joint Program Office.

3. "DOD Programming Language Policy," Linda Brown, Director, STARS Program, Defense Advanced Research Projects Agency.

4. "Army Policy on Ada," Robert Schwenk, Office of the Director for Information Systems and Command, Control, Communications, and Computers, Department of the Army.

5. "Ada in Crisis: The Four Horsemen," Norman Brown, DOD Software Program Managers Network.

6. "Defense Research & Engineering Perspective," Anita Jones, Director, Defense Research & Engineering.

7. "Weapons Program Perspective," RADM K.K. Paige, Technical Director, Navy AEGIS Program.

8. "Ada Policy: Viewpoint," Christine Anderson, Chief, Satellite Control & Simulation Division, Air Force Phillips Laboratory.

9.  "Briefing to NRC Ada Study Committee," Dennis Turner, Director, Software Engineering, Army Communications Electronics Command.

10.  "The Future of Ada," Robert Mathis,  Executive Director, Ada Resource Association.

11.  "Ada:  A Navy Perspective,"  CDR Gary Evans, Naval Information Systems Management Center.

12.  "ASC Perspective on Ada," Phil Babel, Air Force Aeronautical Systems Center.

13.  "Joint Strike Fighter Presentation," CAPT Jules Bartow, USAF, Joint Strike Fighter Program.

14.  "Ada and F-22," COL  Robert Lyons, USAF, Joint Strike Fighter Program.


### Washington, D.C., May 21-23, 1996

1.  "Perspectives on Ada," Jan Lodal, Principal Deputy Undersecretary of Defense  (Policy).

2.  "Transition of Software Projects from Advanced Development," Larry Lynn, Director, Defense Advanced Research Projects Agency.

3.  "Ada," Ike Nassi, Vice President, System Software Technologies, Apple Computer Inc.

4.  "AJPO Rationale," RADM John Gauss, Commander, Joint Interoperability and Engineering Organization and Deputy Director for Engineering and Interoperability, Defense Information Systems Agency.

5.  "Ada Policy," Larry Druffel, South Carolina Research Authority, and John Goodenough, Software Engineering Institute.

6.  "Command and Control Product Lines," and "Effect of Ada on MITRE-Supported ESC Programs," Robert Kent, Director, Software Center, USAF Electronic Systems Center, and Steve Schwarm, MITRE.

7.  "Intentional Programming:  Innovation in the Legacy Age," Charles Simonyi, Chief Architect, Microsoft Research.

8.  "Software at Citicorp," Gerald Pasternack, Citicorp.


### POSITION PAPERS

1.  "AIA Position Statement on Ada," Aerospace Industries Association, July 12, 1996.

2.  "The Ada Mandate," Edward Feigenbaum, Chief Scientist, U.S. Air Force, 1996.

3.  "Comments on Ada Mandate," Huet C. Landry, Defense Information Systems Agency/JEBEB, Software Engineering Standards, 1996.

4.  "Whither Ada?"  Victor Vyssotsky, May 12, 1996.